Robert Boyle

Hydrostatical Paradoxes

Made Out by New Experiments

Robert Boyle

Hydrostatical Paradoxes
Made Out by New Experiments

ISBN/EAN: 9783337712877

Printed in Europe, USA, Canada, Australia, Japan

Cover: Foto ©berggeist007 / pixelio.de

More available books at **www.hansebooks.com**

HYDROSTATICAL

PARADOXES,

Made out by New

EXPERIMENTS,

(For the moſt part

PHYSICAL and Eaſie.)

By the Honourable

ROBERT BOYLE,

Fellow of the Royall Society.

OXFORD,

Printed by *William Hall*, for *Richard Davis*,
Anno Dom. M. DC. LXVI.

·THE

PUBLISHERS

·ADVERTISEMENT·

· TO THE

READE·R·

W Hen the Author Writ the following Trea-tife, he had a .defign, as appears by fome paffages in the Pre-face, to publifh to-gether with it fome things, which he had divers years be-fore provided for an Appendix to his

Phyff;

Phyſico-Mechanical Treatiſe about the Aire : But part of the Appendix conſiſting of Experiments, which the Authour has ſeveral times made, but truſting to his memory, did not think it neceſſary to Record, when he came to recollect particulars, he found that ſome years which had paſs'd ſince divers of them were try'd, and variety of intervening occurrents, had made it unſafe for him to rely abſolutely upon his Memory for all the circumſtances fit to be ſet down in the Hyſtorical part of the deſign'd Appendix. And therefore he reſolv'd to repeat divers Experiments and Obſervations, that he might ſet down their Phænomena whilſt they were freſh in his Memory, if not objects of his ſenſe. But though, when he Writ the following Preface, he did it upon a probable ſuppoſition, that he ſhould ſeaſonably be able to repeat the intended

Tryals,

Tryals, yet his Expectation was fad-
ly difappointed by that heavy, as well
as juft, Vifitation of the Plague which
happened at *London* whil'ft the Author
was in the Country : and which much
earlier then was apprehended, began
to make havock of the People, at fo
fad a rate that not only the Glafs-men
there were fcatter'd, and had, as they
themfelves advertis'd him, put out
their Fires, but alfo Carriers, and o-
ther ways of Commerce (fave by the
Poft) were ftrictly prohibited betwixt
the parts he refided in and *London* ;
which yet was the only place in *Eng-
land* whence he could furnifh himfelf
with peculiarly fhap'd Glaffes, and o-
ther Mechanical Implements requifite
to his purpofes ; And the fame Cala-
mity continuing ftill, without yet af-
fording us any certaine ground of de-
termining when it will end : The Author chufes rather to fuffer the follow-

ing

ing Paradoxes to come abroad without the Appendix, (which is no way neceſſary to them, whatever they may be to It,) then any longer put off thoſe Ingenious perſons that ſolicited the publication of them.

THE

The PREFACE.

He Rife of the following Treatife being a Command impos'd on me by the Royal Society, the Reader will, I hope, need no more then this intimation, to keep him from wondering to find fome paffages worded as parts of a Difcourfe pronounc'd before an Affembly, it being not unufual (though not neceffary) to prefent either in writing or by word of mouth, together with the Experiments made before that Illuftrious Company, an Hiftorical account of Them. But becaufe 'tis probable, that fome Readers will defire to be fatisfied about other particulars, relating to the publication of this Treatife, I prefume it will not be amifs, both to fay fomething of the Rea-

fons,

The Preface.

ſons why I publiſh it as the firſt part of the preſent Appendix to my Phyſico-Mechanical Experiments, and to give ſome account of the manner of writing it.

I had quickly both an opportunity and an Invitation to enlarge the papers I was to read, beyond the Limits of a bare deſcription of the Phænomena, and matters of faɛ̌t, by my having been through ſome intervening Accidents ſo hindered from exhibiting them altogether, that I was deſir'd to bring in an Accompt in Writing, that might be Regiſtred (how little ſoever worthy of ſuch Company) in the Societies Colleɛ̌ion of Philoſophical Papers, for the ſake of thóſe Members who could not be preſent at all the Experiments : So that finding ſome enlargements expeɛ̌ed from me, I was eaſily induc'd to add the Explications of the Phænomena I deſcrib'd, whilſt I perceiv'd that by a ſmall addition of pains I might much gratifie divers Ingenious Friends that were not ſo well vers'd in Hydroſtaticks as in the other parts of real Learning. Ha-

The Preface.

Having thus been induc'd to enlarge the Account of my Experiments till it had attain'd the bulk 'tis now arriv'd at, I confess I was without much difficulty perswaded, that to suffer it to pass abroad * in the Company of the Appendix wherewith 'tis

About this passage, See the Publisher to the Reader.

publish'd, would not prove unacceptable to the Curious, no more then an improper introduction to the rest of my Appendix, and that for several Reasons.

For (first) the Hydrostaticks is a part of Philosophy, which I confess I look upon as one of the ingeniousest Doctrines that belong to it. Theorems and Problems of this Art, being most of them pure and handsome productions of Reason duly exercis'd on attentively consider'd Subjects, and making in them such Discoveries as are not only pleasing, but divers of them surprising, and such as would make one at first wonder by what kind of Ratiocination men came to attain the knowledg of such unobvi-

ous

ous Truths. Nor are the *delightfulness,*
and the *subtilty of the* Hydrostaticks, the
only things for which we may commend
Them : For there are many, as well of the
more familiar, as of the more abstruse Phænomena of Nature, that will never be
throughly understood, nor clearly explicated by those that are strangers to the Hydrostaticks; upon whose Principles depend, besides many other things, the Explications of
most of the Physico-Mechanical Experiments, we have ventur'd to present the
Publick, and the Decision of those many
Controversies, which they, and the Phænomena of the Torrecellian Experiment
have occasion'd among the Modern Inqui-
rers into Nature.

But the use of this Art is not alone
Speculative, but Practical, since not onely
the propositions it teaches, may be of great
importance to Navigation, and to those that
inquire into the magnitudes and gravities
of Bodies, as also to them that deal in Salt
works:

The Preface.

workes: But that the Hydroſtaticks may
be made divers ways ſerviceable to the
Chymiſts themſelves, to whoſe Art
that Doctrine ſeems to be ſo little of Kin,
I might here manifeſt, if I
could think it fit to tranſcribe, what I have * elſewhere deliver'd to that purpoſe.

But that which invited me to Write ſome-
thing of this part of Philoſophy, is, not
only that I think it conſiderable, but that,
notwithſtanding its being ſo, I find it but
very litle, and not very happily cultiva-
ted. For being not look'd upon as a Diſ-
cipline purely Mathematical, the generality of Mathematicians have not in their
Writings ſo much as taken notice of it,
much leſs improv'd It. And ſince the ad-
mirable Archimedes, who, in his little
Tract De inſidentibus humido, has left
us three or four very excellent propoſitions,
(but proved by no very eaſie Demonſtrations)

ons) *among divers others that have more of Geometrical Subtility, then usefulness; Those Mathematicians, that, (like* Marinus Ghetaldus, Stevinus, *and* Galileo) *have added any thing considerable to the Hydrostaticks have been (that I know of) very few, and those too, have been wont to handle them, rather as Geometricians, then as Philosophers, and without referring them to the explication of the Phænomena of Nature. And as for the Peripateticks, and other School-Philosophers, though on some Occasions, as when they tell us, That water weighs not in water, nor aire in aire, they deliver assertions about matters belonging to the Hydrostaticks, (which term, in this Treatise, I often take in a large sense because most of the things delivered about the weight of Bodys may by easy variations, be made applicable to other Fluids) yet they are so far from having illustrated, or improv'd them, that they have but broach'd or*

credited

The Preface.

credited, divers of the most erroneous conceits, that are entertain'd about them. So that, there being but few *Treatises* written about the Hydrostaticks, and those commonly bound up among other *Mathematical* works, and so written, as to require *Mathematical Readers*, this usefull part of *Philosophy*, has been scarce known any farther then by name, to the generality ev'n of those Learned men, that have been inquisitive into the other parts of it, and are deservedly reckoned among the ingenious *Cultivators* of the modern *Philosophy*. But this is not all, For some eminent men, that have of late years, treated of matters Hydrostatical; having been prepossess'd with some erronous Opinions of the peripatetick *School*, and finding it difficult, to consult experience, about the truth of their *Conclusions*, have interwoven divers erroneous *Doctrines* among the sounder *propositions*, which they either borrow'd from *Archimedes*, and other circumspect *Mathematicians*, or devis'd

The Preface.

vis'd themselves, and these mistakes being deliver'd in a Mathematical dress, and mingled with Propositions demonstrably true, the Reputation of such Learned Men, (from which I am far from desiring to detract,) and the unqualifiedness of most Readers, to examine Mathematical things, has procur'd so general an entertainment for those Errors, that now the Hydrostaticks is grown a part of Learning, which 'tis not only difficult to attain, but dangerous to Study.

Wherefore, though neither the Occasion and designe of this Treatise exacted, nor my want of skill and leasure qualified me to Write either a Body or Elements of Hydrostaticks : yet I hop'd I might doe something, both towards the illustrating, and towards the rescue of so valuable a Discipline, by Publishing the ensuing Tract; where I endeavour to disprove the receiv'd errors, by establishing Paradoxes contrary to them, and to make the Truths the

The Preface.

the better understood and receiv'd , partly *by a way of Explicating them unimploy'd in Hydroftatical Books,* and partly *by confirming the things I deliver by Phyfical and fenfible Experiments. And over and above this, the more to recommend Hydroftaticks Themfelves to the Reader, I have, befides the Paradoxes , oppos'd to the Errors I would difprove, taken occafion by the fame way, to make out fome of the ufefulleft of thofe Hydroftatical Truths, that are wont to feem ftrange to Beginners.*

If it be here demanded, why I have made fome of my Explications fo prolix, and have on feveral occafions inculcated fome things. I anfwer, That thofe who are not us'd to read Mathematical Books , are wont to be fo indifpos'd to apprehend things, that muft be explicated by Schemes, and I have found the generality of Learned men, and ev'n of thofe new Philofophers that are not skill'd in Mathematicks , fo

much

much more unacquainted, then I before
imagin'd both with the principles and Theorems of Hydroſtaticks, and with the ways
of explicating and proving them, that
I fear'd, that neither the Paradoxes themſelves, that I maintain, nor the Hypotheſes
about the weight and preſſure of the aire,
upon which, little leſs then my whole Pneumatical Book depends, would be throughly
underſtood without ſuch a clear Explication of ſome Hydroſtatical Theorems, as
to a perſon not vers'd in Mathematical
writings, could ſcarce be ſatisfactorily deliver'd in few words. And therefore,
though I do not doubt, that thoſe who are good
at the moſt compendious ways of demonſtra-
ting, will think, I might in divers places,
have ſpar'd many words without injury to
my proofs, and though I am my ſelf, of
the ſame mind I exſpect to find them of;
yet, I confeſs that 'twas out of choice that
I declind that cloſe and conciſe way of
writing, that in other caſes I am wont moſt

to

The Preface.

to esteeem. For Writing now not to credit my self, but to instruct others, I had rather Geometricians should not commend the shortness of my Proofs, then that those other Readers, whom I chiefly design'd to gratifie, should not throughly apprehend the meaning of them.

But this is not all for which I am to excuse my selfe to Mathematicall Readers. For some of them, I fear, will not like that I should offer for Proofs such Physical Experiments, as do not alwayes demonstrate the things, they would evince, with a Mathematical certainty and accuratenesse; and much less will they approve, that I should annex such Experiments to confirm the Explications, as if Suppositions and Schemes, well reason'd on, were not sufficient to convince any rational man about matters Hydrostaticall.

In

The Preface.

In Anſwer to this. I muſt repreſent, that in Phyſical Enquiries it is often ſufficient that our determinations come very near the matter, though they fall ſhort of a Mathematical Exactneſs. And I chooſe rather to preſume upon the equity of the Reader, then to trouble him, and my ſelf with tedious Circumlocutions, to avoid the poſſibility of being miſunderſtood, or of needing his Candor. And we ſee, that even Mathematicians are wont, without finding any inconvenience thereby, to ſuppoſe all perpendicular Lines, made by pendulous Bodies, to be parallel to one another: Though indeed they are not; ſince, being produc'd, they would meet at the Centre of the Earth: And to preſume, that the Surface of every calme water, in a Veſſel, is parallel to the Horizon; and conſequently, a Plain: Though, in ſtrictneſs, themſelves think it the portion of a Sphere: And though alſo I have uſually

sually observ'd it to be higher, where 'tis almoft contiguous to the fides of the Veffel, then 'tis in other places.

Moreover, fince we find that though water will be uniformly rais'd in Pumps to feveral heights, but not to thirty five foot, and will in ordinary open pipes, be almoft of the fame level within and without, but not if the pipe be extraordinary flender; Upon thefe, and divers other fuch confiderations, I may have fometimes made ufe of ex- preffions, that feemed not pofitive and determinate enough to be employed a- bout matters to which Mathematical De- monftrations are thought applica- ble. But I elfewhere give an account of the fcruples I have about fuch De- monftrations, as they are wont to be ap- ply'd to Phyfical matters. And, in the prefent Paradoxes, I think I have not done nothing, if in my Hydroftatical Ex- plications I have made it appear, That in

Expe-

The Preface.

Experiments made with such Liquors and Glasses, as I employed, the Rules will hold without any sensible, or at least any considerable Error ; for thereby we may learn the Truth of many things, for the main, though in some we should not have attained to the exactness of measures and proportions, which yet our endeavors may assist others to arrive at.

And as for my confirmation of Hydrostatical propositions by Physical Experiments , if some Readers dislike that way , I make no doubt but that the most will not only approve it, but thank me for it. For though, in pure Mathematicks, he that can demonstrate well, may be sure of the Truth of a Conclusion, without consulting Experience about it : Yet because demonstrations are wont to be built upon Suppositions or Postulates ; and some things , though not in Arithmetick or Geometry , yet in Physical matters , are

wont

wont to be taken for granted, about which men are lyable to slip into mistakes ; even when we doubt not of the Ratiocination, we may doubt of the conclusion, because we may, of the Truth of some of the things it supppo-ses. And this Consideration, if there were no other, will, I hope, excuse me to Mathematicians, for ventring to confute some reasonings that are given out for Mathematical demonstrations. For I suppose it will be consider'd, that those whose presum'd Demonstrations I examine, though they were some of them Professours of Mathematicks, yet did not Write meerly as Mathematicians, but partly as Naturalists. : so that to question their Tenets, ought not to disparage those, as well certain, as excellent and most useful Sciences, pure Mathematicks, any more then that the Mathematicians that follow the Ptolemaick, the Copernican, the Tichonian, or other Systemes

of

of the world, Write Books to manifest
one anothers Paralogismes in Astrono-
mical matters : And therefore (to proceed to what I was about to say) it cannot but be a satisfaction to a wary man
to consult sense about those things that
fall under the Cognisance of it, and to
examine by Experiences, whether men
have not been mistaken in their Hypotheses and Reasonings, and therefore the
Learned Stevinus himself (the chief
of the Moderne Writers of Hydrostaticks) thought fit, after the end of his
Hydrostatical Elements, to add in an
Appendix some Pragmatical Examples
(as he calls them) that is, Mechanical
Experiments (how cogent I now inquire
not) to confirm the Truth of his Tenth
Proposition, to which he had, not far
from the beginning of his Book, annexed what he thinks a Mathematical Demonstration. And, about the very Subjects we are now upon, the following
Para-

The Preface.

Paradoxes will discover so many mistakes of eminent Writers, that pretend to have Mathematically demonstrated what they teach, that it cannot but make wary Naturalists (and 'tis chiefly to gratifie such that I publish this) be somewhat diffident of Conclusions, whose proofs they do not well understand. And it cannot but, to such, be of great satisfaction to find the things, that are taught them, verified by the visible testimony of Nature her self. The importance of this Subject, and the frequent Occasion I have to make use of this kind of Apology, will I hope, procure me the Readers pardon if I have insisted somewhat long upon it.

After what has been hitherto discours'd, 'twil be easie for me to give an Account, why I premised these Hydrostatical Paradoxes to the rest of the Appendix, wherewith they are * now publish'd : For since a great part of my work in that Appendix, was to be a further Explication

* An Account of this passage (N6) may be had from the Publishers Advertisement to the Reader.

a 4
of

The Preface.

of some things delivered in the Book it is subjoyn'd to, and the vindication of them from invalid objections : And since I have generally observ'd, that the objections that have bin, either publickly or privately, made against the explications & reasonings contain'd in that Book, were wont to proceed from unacquaintedness, either with the true notion of the weight and spring of the aire, as I maintain them, or with the Principles and Theorems of Hydrostaticks, or else from erroneous Conceits about them ; I thought it would much conduce to both the forementioned ends of my Appendix, If I clear'd up that Doctrine to which my Experiments and reasonings have been all along Consonant, & whose being either not known, or misunderstood, seems to have occasion'd the objections that have been hitherto made against the Hypotheses I have propos'd, or the Explications I have thence given. And however, since the Proofs I offer for my opinions are for the most part drawn from

Expe-

The Preface.

Experiments new & eafie, and that my aim is but to difcover Truths, or make them out by clearer explications, without fuppofing, like thofe I diffent from, any thing that is either precarious *or* fcarce, *if at all,* intelligible; *I hope, that if I fhould not prove happy enough to reach my ends, yet the Ingenious and Equitable Reader will approve my Defigne, and be advantaged by my Experiments. Of which fome of the chiefeft, and fome of the moft difficult, having been feen (divers of them more then once) by the* Royal Society *it felf, or by inquifitive Members of it; it will, I prefume, be but a reafonable requeft, if the Reader, that fhall have the curiofity to try them over again, be defired not to be hafty in diftrufting the matters of faEt, in cafe he fhould not be able at firft to make every thing fucceed according to expeEtation. For as eafie as I have endeavour'd to make thefe Experiments, yet I dare not promife my felf that they will all of them be priviledg'd from the fate*

where-

The Preface.

whereto I have observ'd other Physico-Mathematical ones to be not seldome obnoxious from some unheeded Physical Circumstance, by which those that are not acquainted with the subtleties of Nature, or, at least for the time, do not sufficiently consider them, are apt to be imposed upon.

This Advertisement will perhaps be best illustrated, & recommended by an instance. And therefore I shall subjoyne one that will possibly seem somewhat odd.

It has been taken notice of by two or three Ingenious modern Mathematicians, and I have had occasion to make it out by particular Experiments, that warm water is lighter in specie then cold : whence it has been deduc'd, that wax, and other Bodies, very near aequiponderant with common water, will swim in that which is cold, and sinck in that which is hot, or luke-warm. Which Experiment, though as it may be (and perhaps it has been) tryed, I readily allow to be agreeable to the known Laws of the Hydrostaticks ;

Yet

Yet I have sometimes undertaken that the Tryal should have a quite contrary event. To this purpose having taken some yellow Bees-wax, which was formed into a Pellet of the bigness of a Cherry, and, by the help of a little Lead, was made so near æquiponderant to cold water, that, being but a very little heavier, a very small diminution of its weight would make it emerge, I remov'd it out of the very cold water, into some that had bin purposely made lukewarm, (or a little more then so) where it quickly, somewhat to the wonder of the lookers on, appeard to swim on the top of the water. And that it might not be suspected that it was supported by any visible bubbles, which I have observed, in some cases, to buoy up even heavy Bodies, and deceive the unskilful, or unattentive; I briskly enough duck'd the bullet 2 or 3 times under water to throw them off, notwithstanding which it constantly return'd to float, and yet being remov'd again into the same cold water it had been taken

out

out of, and duck'd as before to free it from
adherent bubbles, it lay quietly at the bottom, and, though rais'd several times to the
upper part of the water, would immediately
subside again, and fall to the very lowest.
Now that wch invited me to promise an Experiment which seems to contradict the principles of the Hydrostaticks, was not any di-
strust of those principles themselves, but a
conjecture, that as by warmth the water
would be made a little lighter in specie then
'twas before; so by the same warmth the spi-
rituous and more agitable parts of the wax,
whose texture is loose enough, would be som-
what (though not visibly) expanded, and
would by that expansion gain a greater ad-
vantage towards floating, then the increas'd
lightness of the water would give it dispofi-
tion to sinck. And I confirm'd this conjecture
by a farther experiment, which at first was
it self somewhat surprising to the Beholders.
For when the wax was first taken out of the
cold water, & immediately immers'd in the
warm,

warm, it would readily enough finck, & being
(with a quill or a knife) rais'd to the top of
the water, it would again fall down, but more
flowly then at the begining, & after some few
minutes, if it were rais'd to the upper parts
of the water, it would remain a float. (And I
have known it, when it had remain'd a while
longer at the bottom, so to emerge, that if I
were sure no unheeded bubbles had been
newly generated, and held it up, it might be
said to emerge of its own accord) as on the o-
ther side, being put into the cold water as soon
as ever it was taken out of the warm, it would
at the very first float, and being then knock'd
downwards, it would, readily enough, regain
the upper part of the water, but if I con-
tinu'd to send it downwards about 6 or 7
times (more or fewer) succeffively, it would
emerge every time more flowly then other,
and at length not emerge at all, even when I
try'd it in water made heavy, by being highly infrigidated with falt aud fnow plac'd a-
bout the Glaſs. Which Phænomena I had
thought

thought it reasonable to expect, because I
presum'd, that the Wax being remov'd immediately out of the warm water, into the
cold, must require some time, to loose the
adventitious expansion, which the warmth
had given it, and must be depriv'd of it by
degrees, by the coldness of the water into
which the wax was transferr'd. As on the
other side, there must be some time necessary
for so little a warmth, as that of the tepid (or
little more then tepid) water, to give the
wax that addition of dimensions (which also
it must receive by degrees) that was necessary, in spite of the rarefaction of the water, to make it float. I might add, that these
Tryals were repeated, for the main, with
more Bullets of wax then one, and that they
succeeded far otherwise, when, instead of a
piece of wax, we imploy'd a pois'd glass
bubble, in which the temperature could make
either no change at all, or no considerable
change of dimensions. And to these I might
add other circumstances, if I did not remember,

ber, that I mention these Tryals but occasi-
onally, and to make the caution, formerly recommended to the Reader, appear not to be
impertinent, since a Hydrostatical Experiment true in its self, may easily miscarry by
over-looking such Circumstances as 'tis not
easie to be aware of.

But by this Advertisement I would by
no means divert Men from being diffident of
Hydrostatical Traditions and Experiments.
For, besides the many Erroneous Opinions,
there are matters of fact, whose Truth, thô
not question'd, but built upon, I think ought
to be brought to tryal. For, even whilst I was
concluding this Preface, I found that divers
even of the Moderns, & particularly a very
learned Man that has lately Writen of Hydrostaticks, have much troubled themselves
to render a reason why, since, according to
their Doctrine, water weighs not in water,
Wooden vessels, though of a substance lighter then water, being by leaks, or otherwise,
fil'd with water, should sinck and remain at
the

the bottom of the water: whereas judging this Phænomenon difagreable to what I look upon as the Laws of the Hydroftaticks, I was confirm'd in that opinion, by having had the curiofity to make fome tryals of it, with 4 or 5 veffels of differing fhapes and fizes, whereof two were of wax, which, though a matter but very little lighter then water, I could not finck, or keep funck by pouring water into them, or fuffering them to fill themfelves at leaks made near the bottom, and if they were depreffed by force or weights, they, as alfo the wooden Veffels, would upon the removal of the impediment (and fometimes with the cavity upwards) emerge. And I am the more folicitous to have things in the Hydroftaticks duly af-certain'd, becaufe the weighing of bodies in Liquors may hereafter appear to be one of the general ways I have employ'd, and would recommend, for the examining of al-moft all forts of tangible Bodies.

HYDRO-

THE CONTENTS.

qual

THE CONTENTS.

HYDRO-

Imprimatur

ROBERTUS SAR...

Vicar.-General...

Imprimatur,

ROBERTUS SAY,

VICE-CANCELLARIUS

OXON.

❦❦❦❦❦❦❦❦❦❦❦❦❦❦❦❦❦❦❦❦❦❦❦❦❦❦

HYDROSTATICAL
PARADOXES,
Made out by
NEW EXPERIMENTS:
Prefented to the
ROYAL SOCIETY;
(*The* Lord Vifcount Brouncker *being*
then Prefident.) May 1664.

My LORD,

T O obey the orders of the Society, that forbid the making of Prefaces and Apologies in Accounts of the Nature of that which you expect from me ; I fhall without any further preamble begin with taking notice, that

B   upon

upon perusal of *Monsieur Paschall's* small French Book , which was put into my hands , I find it to consist of two distinct Treatises : The one *of the Equilibrium* of Liquors, as he calls it ; and the other *of the weight of the Mass of the Air.*

: As for this latter, (which I shall mention first , because I can in very few words dispatch the little I have to say of it,) Though it be an ingenious discourse , and containes things, which if they had been published at the time , when it is said to have been written , would probably have been very well-come to the Curious : yet I have very little else to say of it in this place , in regard that since that time , such kind of Experiments have been so prosecuted, that I presume it is needless , and would not be acceptable to repeat what *Monsieur Paschall* has written, in this Society ; which has seen the same

Truths

Truths, and divers others of the like
Nature, more clearly made out by Experiments, which could not be made
by *Monſieur Paſchall*, and thoſe other
Learned Men, that wanted the advan-
tage of ſuch Engines and Inſtruments,
as have in this place been frequently
made uſe of.

Wherefore having already at a former meeting given you, by word of
Mouth, an account of *Monſieur Paſchall*'s
Ingenious Invention, of a pair of Bel-
lows without vent, to meaſure the
various Preſſure of the Atmoſphære ;
I remember nothing elſe that needs
- hinder me from proceeding to the other
part of his Book, *The Treatiſe of the
Æquilibrium of Liquors*.

This I find ſo ſhort, and ſo worthy
of the Author, that to give you all that
I judge worth taking notice of in it,
would obliege me to tranſcribe *almoſt*
the whole Tract ; and therefore I ſhall
rather

rather invite you to read the whole,
then divert you from the defigne by
culling out any part of it; yet if you
will not be fatisfied without fomething
of more particular, I fhall be oblig'd
to tell you, That the Difcourfe confifting partly of Conclufions and partly
of Experiments; the former feemed to
me to be almoft all of them (there being but few that I doubt of) confonant
to the Principles and Lawes of the
Hydroftaticks. But as for the latter,
the Experimental proofs he offers of his
opinions are fuch, that I confefs I
have no mind to make ufe of them.

And the Reafons why, notwithftanding that I like moft of *Monfieur Pafchall's* Affertions, I decline imploying
his way of proving them, are principally thefe.

Firft, Becaufe though the Experiments he mentions be delivered in fuch
a manner, as is ufual in mentioning
matters

(5)

matters of fact; yet I remember not
that he exprefly fays that he actually
try'd them, and therefore he might
poffibly have fet them down as things
that *muft* happen, upon a juft confidence that he was not miftaken in his
Ratiocinations. And of the reafona-
blenefs of this Doubt of mine, I fhall
ere long have occafion to give an
inftance.

Secondly, Whether or no *Monfieur*
Pafchall ever made thefe Experiments
himfelf; he does not feem to have been
very defirous, that others fhould make
them after him. For he fuppofes the
Phænomena he builds upon to be produc'd fifteen or twenty foot under water. And one of them requires, that
a Man fhould fit there with the End
of a Tube leaning upon his Thigh.
But he neither teaches us how a Man
fhall be enabled to continue under water, nor how in a great Ciftern full of

B 3

water

water, twenty foot deep, the Experi-
menter shall be able to discern the al-
terations, that happen to *Mercury* and
other Bodies at the Bottome.

And *Thirdly*, These Experiments
require not only Tubes twenty foot
long, and a great Vessel of at least as
many feet in depth, which will not in
this Countrey be easily procured, but
they require Brass Cylinders, or Pluggs,
made with an exactness, that, though
easily supposed by a Mathematician,
will scarce be found obtainable from a
Tradesman.

These difficulties making the Experiments propos'd by *Monsieur Paschall*
more ingenious then practicable, I was
induc'd on this occasion to bethink my
self of a far more Expeditious Way,
to make out, not only most of the
Conclusions wherein we agree, but o-
thers that he mentions not; and this
with so much more ease and clearnesse,
That

(7)

That not only This Illuſtrious Aſſem-
bly, but perſons no more than mode-
rately vers'd in the Vulgar principles
of the Hydroſtaticks, may eaſily e-
nough apprehend what is deſign'd to
be deliver'd; if they will but bring
with them a due Attention, and
minds diſpos'd to preferre Reaſon and
Experience to vulgar Opinions and
Authors; which laſt clauſe I annex,
becauſe the following Diſcourſe, pre-
tending to confute ſeveral of thoſe, chal-
lenges a right to except againſt their
Authority.

It not being my preſent Task to deliver the Elements, or a Body of Hydroſtaticks, but only ten or twelve Paradoxes, which I conceive to be prove-
able by this new way of making them
out, I ſhall, to avoid Confuſion, Deliver Them in as many diſtinct propoſitions; After each of which, I
ſhall indeavour in a proof, or an Explication,

B 4

plication, to show, both that it is true, and why it ought to be so. To all these I shall to avoid needless Repetitions, premise a word or two by way either of *postulatum* or *Lemma*.

And because I remember to what Assembly I address This Discourse, I shall make use of no other then an easie supposition I met with in a short Paper (about a Mercuriall Phænomenon) brought in a year or two since to this Learned Society, by a deservedly Famous Member of it *, For though his supposal be made upon occasion of an Experiment of another Nature, then any of the ensuing, it may be easily accomodated to my present purpose.

* *That excellent Mathematician the Learned Dr* Wallis, Savilian Professor of Geometry.

This *postulatum* or *Lemma*, consists of three parts; the first of them more, and the two last, less principal.

Suppose

Suppofe we then, (*Firſt*) That if a
Pipe open at both Ends, and held perpendicular to the Horizon, have the
lower of them under Water, there
paſſes an Imaginary plain or Surface,
which touching that Orifice is parallel
to the Horizon ; and confequently parallel as to fenfe to the upper Surface of
the water, and this being but a help to
the Imagination will readily be granted.

Secondly, To this it will be confonant, that each part of this defignable
furface, will be as much, and no more
prefs'd, as any other equal part of it,
by the water that is perpendicularly
incumbent on it. For the water or
other Fluid being fuppofed to be of an
homogeneous fubſtance, as to gravity,
and being of an equal height upon all
the parts of the imaginary Surface ;
there is no reafon why one part ſhould
be more prefs'd by a perpendicular pillar of that incumbent fluid, then any
other

other equal part of the fame Surface
by another perpendicularly incumbent
pillar of the fame or equal Bafis and
height, as well as of the fame Liquor.

But *Thirdly*, Though whilft our
imaginary Surface is equally prefs'd upon in all parts of it, the Liquor muft
retain its former pofition; yet if any
one part comes to have a greater weight
incumbent on it, then there is upon
the reft, that part muft be difplac'd,
or deprefs'd, as it happens, when a
ftone or other Body heavier then water
fincks in water. For whereyer fuch a
a Body happens to be underneath the
water, that part of the imaginary plain
that is contiguous to the lower part of
the ftone, having on it a greater weight
then other parts of the fame Surface,
muft needs give way, and this will be
done fuccefively till the ftone arrive at
the Bottom; and if, on the other fide,
any part of the Imaginary Surface be

lefs

lefs prefs'd upon then all the reft ; it
will by the greater preflure on the other
parts of the Surface be impell'd upwards, till it have attain'd a height, at
which the preflure (of the rais'd water,
and the lighter or floating Body (if any
there be) that leans upon it, and gravitates together with it, upon the
fubjacent part of the Imaginary Surface) will be equal to that which
bears upon the other parts of the fame
Surface.

And becaufe this feems to be the
likelieft thing to be Queftion'd in our
Affumption, though he
that confiders it atten-
tively, will eafily enough
be induc'd to grant it :
Yet I fhall here endea-
vour to evince it Expe-
rimentally, and that
by no other way of proof, then the
fame I imploy all along this prefent difcourfe.   Take

*This Experiment and
the Explication of it,
if to fome they fhould
here 'feem fomewhat
obfcure, will be eafily
underftood by the Fi-
gures and Explicati-
ons belonging to the
firft enfuing Para-
doxe.*

Take then a Cylindrical glafs pipe, of a convenient Bore open at both Ends, let the Tube be fteadily held perpendicular to the Horizon, the lower end of it being two or three inches beneath the Surface of a convenient quantity of water, which ought not to fill the Glafs Veffel that contains it. The pipe being held in this pofture, 'tis manifeft, that the water within the pipe, will be *almoft* in a level with the Surface of the water without the pipe, becaufe the external and internal water (as I am wont for Brevities fake to call them) have free intercourfe with one another by the open Orifice of the immers'd End of the pipe : yet I thought fit to infert the word *almoft*, becaufe if the pipe be any thing flender, the Surface of the water in it, will always be fomewhat higher then that of the water without it, for reafons that 'tis not fo neceffary we fhould now inquire after,

after ", as 'tis , that we should here de-
sire to have this taken notice of once for
all ; That miftakes may be avoided
without a troublefome repetition of the
difference in heights of the Surface of
Liquors within pipes and without
them , in cafe they be any thing
flender.

The pipe being held in the newly
mention'd pofture , if you gently poure
a convenient Quantity of Oyle upon
the external water , you fhall fee, That
as the Oyle grows higher and higher
above the Surface of That water , the
water within it , will rife higher and
higher , and continue to do fo , as long
as you continue to poure on oyle ; Of
which the Reafon feems manifeftly to
be this ; That in the Imaginary plaine
that paffes by the Orifice of the immers'd end of the pipe , all that is not
within the Compafs of the Orifice , is
expos'd to an additional preffure from
the

the weight of the oyle which swims
upon the water , and that pressure must
still be increas'd , as there is more and
more oyle poured on ; whereas a Circular part of the Imaginary plain, equal
to the Orifice of the Glasse , is by
the sides of the pipe fenc'd from the
immediate pressure of the oyle ; so that
all those other parts of the water , being far more press'd , then that part
which is comprehended within the Cavity of the Tube : and consequently the
press'd parts of the external water, are
by the equal gravitation of the oyle ,
upon the parts of the external water ,
impell'd up into the Cavity of the pipe,
where they find less resistance , then
any where else ; till they arrive at such
a height , that the Cylinder of water,
within the pipe , do's as much gravitate upon the subjacent part of the Imaginary Surface , as the water and
oyle together , do upon every other
equal

equal part of the same Surface or plain.
But as well the former *Lemma*, as this
Experiment, will be sufficiently both
clear'd and confirm'd by the following
Explications; to which I should for
that Reason forthwith proceed; Were
it not that, since divers passages of the
following Treatise, suppose the Aire
to be a Body not devoid of weight,
which yet divers Learned adherents to
the Peripatetick Philosophy do reso-
lutely deny, it seems requisite to premise something for the proof of this
Truth.

And though I think the Arguments
we have imploy'd to that purpose already, do strongly evince it: yet if I may
be allow'd to anticipate one of my own
Experiments of the Appendix, I shall
give an instance of the weight of the
Aire, not lyable so much as to those
invalid objections, which *some of* the
Aristotelians have made against those
Proofs

Proofs, wherewith we have been fo happy, as to fatisfie the learned'ft even of our profeffed Adverfaries.

We caus'd then to be blown at the flame of a Lamp, a Bubble of glafs, (of about the bignefs of a fmall Hen-egge) which, that it might be light enough to be weigh'd in exact Scales, ought to be of no greater thicknefs, then is judged neceffary to keep it from being (when feal'd up with none but ve-ry much expanded aire in it) broken by the preffure of the ambient Atmo-fphære. This bubble was (like a Peare with its ftemme) furnifh'd with a very flender pipe of Glafs, at which it was blown, that it might be readily feald up; and then (the Aire within it being by the flame of the Lamp gradu-ally rarified, as much as conveniently could be) whilft the Body of the Bub-ble was exceeding hot, the newly mentioned ftemme was nimbly put into the

the middle of the flame; where, by
reafon of its flendernefs, the Glafs,
which was exceeding thin, was immediately melted; whereby the Bubble
was Hermetically feal'd up. This
Glafs being permitted leafurely to
coole, I could afterwards keep it by
me an hour, or a day, or a week, or
longer, if I thought fit; and when I
had a mind to fhew the Experiment,
I put it in one of the fcales of an exact
ballance, that would turn, perhaps
with the 30*th*, or 50*th*, or a leffe part of
a grain; and having carefully counter-
pois'd it, *I* then warily broke off the
feal'd end, placing a fheet of paper juft
under the fcale to receive the fragments
of the Glafs : and putting in again
thofe fragments, that fcale wherein
the Glafs was would confiderably preponderate; which it muft do upon
the account of the Weight of Aire,
there being no other caufe, either

C   needful,

needful , or juftly affignable , but the
weight of the Aire that rufh'd into the
Cavity of the Glafs , as finding lefs
refiftance there then elfewhere , by
reafon that the included Aire had it's
fpring much weakn'd by it's great ex-
panfion.

This Experiment I many times tryed , fometimes before fome *Virtuofi*,
and fometimes before others ; who all
allowed it to be conclufive. For here
it could not be objected as againft the
weighing of Aire in a Bladder , (which
objections yet I could eafily anfwer ,
if it were *now* proper) that the aire
which ponderates , is ftuff'd with the
Effluvia of him that blows the Bladder , and (befides that) is not aire in
its Natural ftate, but violently comprefs'd. For here 'tis the free aire, and
in it's wonted laxity, that makes the
Glafs preponderate.

And that there is a great Ingrefs of
the

the external aire, is evident by thefe
three Phænomena. The *one*, that if you
lend an attentive Ear, you fnall plainly
heare a kind of whiſtling noiſe to be
made by the external aire, as it ruſhes
violently in upon the breaking of the
Glaſs; The *other*, that the Rarefaction of the aire, feal'd up in the bubble, being very great, there is a great
deal of fpace left for the ambient aire
to fill upon its admiſſion; and the great-
neſs of this Rarefaction may be gueſs'd
at, both by the breaking of fuch bubbles now and then by the preſſure of the
External aire, which is not competently affifted by the Internal to refiſt; and
alfo by the *third* Phænomenon I intended to take notice of, namely, That if,
inſtead of breaking off the feal'd end of
the Glaſs in the aire, you break it under water, that Liquor will, by the
Preſſure of the Atmoſphære, be forc'd
to fpring up like an artificial Fountaine
into

into the Cavity of the Bubble, and fill about three quarters of it. By which laſt circumſtance I gather, that the weight of the aire is more conſiderable then ev'n many, who admit the aire to have weight, ſeem to imagine. For we muſt not ſuppoſe, that all the aire contain'd in the Bubble, when broken, weighs no more then the weight requiſite in the oppoſite Scale, to reduce the Ballance to an Æquilibrium; ſince this additional weight is onely that of the aire, that intrudes on the breaking of the glaſs; which aire, by the Obſervations newly mention'd to have been made with water, appears to be but about three quarters of the whole aire contain'd in the broken Bubble; and yet, according both to our Eſtimate, and that of divers Virtuoſi, and ſome of them eminent Mathematicians, when the capacity of the Bubble was ſhort of two cubical Inches, (and ſo proportionably

tionably in other glaſſes,) the nice Ballance we us'd, manifeſted the newly admitted Aire to amount to ſome times
near halfe a grain, and ſometimes beyond it.

　And becauſe one of the laſt Experiments that I made to this purpoſe,
with ſeal'd Bubbles was none of the
leaſt accurate, I ſhall conclude this
Subjeƈt with the following account
of it.

　A thin glaſs Bubble, blown at the
flame of a Lamp, and Hermetically
ſeal'd when the contained aire was exceedingly rarified, was Counterpoiz'd
in a nice paire of Scales, and then the
ſeal'd *apex* being broken off, and put a-
gain into the ſame Scale, the weight appear'd to be increas'd by the re-admit-
ted aire, a pretty deal above $\frac{11}{16}$ᵗʰˢ, and
conſequently very near, if not full ¾ of
a graine : Laſtly, having by ſome ſlight
(for 'tis no very eaſie matter) fill'd it

with

with common water, we weigh'd the
glaſs and water together, and found the
latter, beſides the former, to amount
to 906 grains :. ſo that ſuppoſing, according to our former Eſtimate, coun-
tenanced by ſome Tryals, that the re-
admitted aire, which amounted to $\frac{3}{4}$ of a
grain, fill'd but $\frac{3}{4}$ of the whole Cavity of
the Bubble, the aire that was in it,
when ſeal'd, poſſeſſing one quarter of
that Cavity, the whole aire contain'd
in the Bubble, may be reaſonably pre-
ſum'd to weigh a whole grain; in which
caſe we might conclude (abſtracting
from ſome little Niceties not fit to be
taken notice of here, as elſewhere) that
the water in our Experiment, weighed
very little more then nine hundred-times
as much as an equal quantity of Aire,
And therefore, though we allow, that
in an Experiment ſo diligently made, as
this was, the aire præexiſtent in the bubble did not adæquately poſſeſs ſo much
as

as a fourth part, but about a fifth or a
sixth of its Cavity, the aire will yet appear so heavy, that this Experiment will
agree well with thofe others, recorded
in another Treatife, wherein we affign'd
(*numero rotundo*) a thoufand to one, for
the proportion wherein the fpecifick
Gravity of water exceeds that of aire.

PARA-

PARADOX I.

That in Water, and other Fluids, the low-er parts are pres's'd by the upper.

PRovide a Glass vessel of a convenient height and breadth A. B. C. D.
fill'd with water almost to the Top;
Then take a glass Pipe, open at both
Ends, Cylindrical, and of a small Bore,
(as about the eighth or sixth part of an
Inch in Diameter.) Put the lower End
of this Pipe into clear Oyle or Spirit
of Turpentine; and having by Suction
rais'd the Liquor to what part of the
Pipe you think fit, as soon as it is there,
you must, very nimbly removing your
Lips, stop the upper Orifice with the
pulp of your finger, that the rais'd Liquor may not fall back again: Then
taking

taking the Pipe and that Liquor out of
the Oyle of Turpentine, place it perpendicularly in the Glafs of water, fo
as that the Surface of the Oyle in the
Pipe be fomewhat higher then that of
the water without the Pipe ; and having
fo done, though you take off your finger from the upper Orifice of the Pipe,
the Oyle will not fall down at the lower
Orifice, though that be open, but will
remain fufpended at the fame height, or
near there abouts, that it refted at
before.

Now Oyle of Turpentine, being a
heavy Fluid, does, as fuch, tend downwards, and not being ftopp'd by the
Glafs it felf, whofe lower Orifice is
left open, it would certainly fall down
through the Pipe, if it were not kept fufpended by the preffure (upwards) of the
water beneath it. There appearing no
other Caufe to which the Effect can
reafonably be afcrib'd, and this being
fuffici-

sufficient to give an Account of it, as we shall presently see. For that it is not any contrariety in Nature, betwixt the oyle and the water, as Liquors that will not mingle, is evident from hence, That if you had remov'd your finger when the Pipe was not so deeply immers'd in the Glass, but that the Surface of the oyl in the Pipe was an Inch or two more elevated above that of the water in the Glass, then in our present case we suppose it to be ; The Oyle, notwithstanding its presum'd contrariety to water, would have freely subsided in the Pipe, till it had attain'd an æquipollency of pressure with the External Water.

The Reason therefore of the Phænomenon seems to be plainly this. Supposing the imaginary surface, on which the Extremity Q of the pipe $P\,Q$ leans, to be $G\,H$. If that part of the Surface, on which the Oyl leans at Q, be as much, and no more charged, or press'd upon

by

by the weight of the incumbent Cylinder of Oyle $Q\,X$, then the other parts
of the same imaginary Surface $G\,H$
are by the water incumbent on Them,
there is no Reason why that part at Q
should be displac'd, either by being depress'd by the weight of the Cylinder of
Oyle $X\,Q$, or rais'd by the equal pressure of water upon the other parts of
the Superficies $G\,H$.

And that this *Æquilibrium*, betwixt
the Oyle and the Water, is the true
cause of the Phænomenon, may be confirm'd by observing what happens, if
the altitude of either of the two Liquors
be alter'd in Relation to the other.

And (*First*,) we have already taken
notice, That if the Cylinder of Oyle
reach in the Pipe, much higher then
that of the Surface of the water, the
oyle will descend : Of which the Reason is, Because the designable Surface
$G\,H$, being more charg'd at Q then
any

any where elfe, the part $\mathcal{Q}$, being unable to refift fo great a preflure, muft ne-
ceflarily be thruft out of place by the
defcending oyle.

Secondly, This fubfiding will continue but till the Surface of the Oyle in
the Pipe be fallen almoft as low as that
of the water without the Pipe ; becaufe
then, and not before, the parts at $\mathcal{Q}$ are
but as much prefs'd by the oyle, as the
other parts of the Surface G H are
by the water that leans upon them.

Thirdly, 'Tis a concluding Circumftance to our prefent purpofe, That if
the Oyle and Water being in an *Æqui-*
librium, you gently lift up the Pipe, as
from $\mathcal{Q}$ to S, the depth of the water
being leflend, the oyle in the Pipe will
grow præponderant, and therefore will
fall out in Drops or Globuls, which by
the greater Specifick Gravity of the
water, will be buoy'd up to the Top
of the Liquor, and there flote : And
ftill

ftill as you lift up the Pipe higher and
higher, towards the Surface *L M*, more
and more of the Oyle will run out.
But if you ftop the Pipe any where in
its Afcent, as at *S*, the Effluxion of the
oyle will likewife be ftopp'd. And at
the imaginary Superficies *J K*, as by
Reafon of the fhallownefs of the water
from *L* to *J*, or *M* to *K*, the preffure
of the water upon the other parts of
the Surface is not near fo great, as it
was upon the Surface *G H*, where the
water had a greater depth : So by reafon of the proportionate Effluxion of
the oyle, whil'ft the Pipe was lifted up
from *Q* to *S*, the remaining Cylinder
of oyle incumbent on *S*, is not able
to prefs that part of the Superficies
J K more ftrongly then the other parts
of the fame Superficies, are preft by the
water Incumbent on them. And if the
Pipe be lifted up till the lower Orifice
be almoft rais'd to *V*; that is, almoft

as

as high as the uppermost Surface of the
water *L. M*, so much of the oyle will,
for the Reason already given, run out,
that there will scarce be any left in the
Pipe *T V*.

Fourthly, But if when the Pipe rests
at the Surface *G H*, where the oyle is
in an *Æquilibrium* with the water, you
should instead of lifting it from *Q* to *S*,
thrust it down from *Q* to *O*; then the
External water would not only sustaine
the oyle, but make it ascend in the Pipe
to a height equal to the distance *E·G*;
and so the Pipe will containe besides a
longer Cylinder of oyle *Æ W*, a shorter one of water *Æ O*. For the pipe
being transferr'd from the position *P Q*,
to the position *O N*, there is a new I-
maginary Surface *E·F*, that passes by
the lower Orifice of the Pipe. Now
the part of this Surface at *O* will not,
by the Incumbent oyle alone, be press'd
as much as the other parts of the same
Sur-

Surface are by the Incumbent water.
For the oyl alone was but in *Æquilibrium*
with the water, when it was no deeper
then *L G*, or *H M*; fo that the other
parts of the Superficies *E F*, being
more prefs'd upon by the water, then
the part at *O* by the oyle, the oyle muft
give place, and be buoy'd up by the
water, (which, if it were not for the
weight of the oyle, would be impell'd
up into the pipe full as high as the Surface of the External water) till the preffure of the admitted water *O Æ*, and
the Cylinder of oyle *Æ W*, do both together gravitate as much upon the part
O, as the reft of the Incumbent water
does upon the other parts of the fame
Superficies *E F*.

Fifthly and laftly, 'Tis very agreeable to what has been delivered, touching the *Æquilibrium* of the oyle and
water in the pipe *P Q*, that the Surface *X* of the oyle in the pipe, will not
be

be of the fame level with *L M*, that
of the External water, but a little
higher than it. For though the flendernefs of the Pipe do fomewhat contribute to this Effect, yet there would
be an inequality, though not fo great,
betwixt thefe Surfaces upon this Account, That oyle of Turpentine being
im*Specie*, (as they fpeak in the Schools)
that is bulk for bulk, a lighter Liquor
than Water, it is requifite that the
height of it, incumbent on the part Q,
be greater than that of the water on the
other parts of the fame Surface *G H*,
to make the preffure of the oyle on the
part it leans upon, equal to the preffure of the water on the other parts of
the Surface. And if the inequality
were greater betwixt the Specifick
Gravities of thefe two Liquors, the
inequalities betwixt the Surface *X*, and
the Surface *L M* would be alfo greater,
as may be try'd by fubftituting for common

mon water, oyle of Tartar *per deliqui-um*, which is a faline Liquor much heavier than it. And that, in cafe the Pipe containe not a lighter Liquor then the External fluid, the Surface of the Liquor in the Pipe will not be higher than that of the Liquor without it, we fhall by and by have opportunity to manifeft by Experience.

From what has been hitherto fhewen, we may fafely infer the Propofition, upon whofe occafion all this has been delivered. For fince the oyle in a Pipe, open at both Ends, may be kept fufpended in any part under water, as at Q, becaufe it is there in an *Æquilibrium* with the External water; and fince being lifted up in the water, as from Q to S, the oyle can no longer be kept fufpended, but by its own gravity will runne out. And fince, in a word, the deeper the water is, the greater weight and preffure is requir'd in the Cylinder of

D   oyle,

oyle, to be able to countervail the preffure of the water, and keep it felf from
being lifted up thereby ; there feems
no caufe to doubt but that the parts of
the water incumbent on the Superficies
G H, do more prefs that Superficies,
than the parts of the water contiguous
to the Superficies *J K* do prefs that ;
and confequently, that the parts of the
water that are under the uppermoft
Surface of it, are prefs'd by thofe of
the fame Fluid that are directly over
them : As we faw alfo that the upper
parts of the oyle, whil'ft the pipe was in
raifing from *Q* to *S.*, deprefs'd the lower fo much, as to force them quite out
of the Pipe ; there being in thefe cafes
no reafon why the lowermoft parts of
a Liquor fhould prefs more, or have a
ftronger Endeavour againft any other
Liquor (or any other Body) the higher
the Liquor incumbent reaches, if thefe
inferiour parts deriv'd their peffure only

ly from their own particular Gravity, (which is no greater then that of the other Homogeneous parts of the Liquor)and therefore they muſt derive the great force wherewith they preſs from the weight of the Incumbent parts, which conſequently muſt be allow'd to preſs upon them.

But before I proceed to the following propoſitions, it will not be amiſs to mention here, once for all, a few advertiſements, to avoid the neceſſity of repeating the ſame things in the ſequel of the Diſcourſe.

And *Firſt,* What is here ſaid of the preſſure of the parts of water upon one another, and the other Affections that we ſhall attribute to it, in the following paper, are to be apply'd to heavy Fluids in general, unleſs there ſhall appear ſome particular Cauſe of excepting ſome of them in particular Caſes.

Secondly, Whereas I lately intima-

ted,

ted, That the inequality betwixt the
Surfaces of the oyle in the Pipe, and of
the External water, was in part to be
afcrib'd to the flendernefs of the Pipe,
to be imploy'd in thefe Experiments,
I did it for this caufe, that, whatever the
Reafon of it be, (which we need not
here inquire after,) we are affur'd by
Experience, as we have elfewhere
fhewn, That when Glafs pipes come
to be flender, water and many other
Liquors (though not Quickfilver) will
have within them a higher Surface then
that of the fame Liquor without them,
and this inequality of Surfaces (as far
as we have yet try'd) increafes with the
flendernefs of the pipe. But this, as to
our prefent Experiment, is a matter of
fo little moment, That it may fuffice
to have intimated that we did not over-
fee it.

Thirdly, Wherefore, notwithftanding this little inconvenience of flender
Glaffes,

Glaſſes, we think it Expedient to imploy ſuch in the following Experiments, becauſe we found, that in thoſe
of a wide Bore, upon ſuch little inequalities of preſſure as are not eaſily to
be avoided, the oyle and water will paſs
by one another in the Cavity of the
pipe, and ſo ſpoile the Experiment,
which requires that the oyle within the
pipe be kept in an intire and diſtinct
Body.   :

Fourthly, Common oyle and water,
or any other two Liquors that will not
mingle, may ſerve the turn in moſt of
theſe Experiments; but we rather chuſe
oyle of Turpentine, becauſe it is light
and thin, clear and colourleſs, and may
be eaſily had in quantities, and is not
ſo apt to ſpot ones Cloaths, or obſti-
nately to adhere to the porous Bodies it
chances to fall on, as Common, and o-
ther expreſs'd oyles.   And for their
ſakes to whom the odour is offenſive,

we

we prefently correct it, by mingling
with it a convenient quantity of oyle
of Rhodium, or fome other Chymical
oyle that is odoriferous.

Fifthly, Oyle of Turpentine, though
it be not reckon'd among the faline
Menftruums, will yet (as we elfewhere
note) work upon Copper, and fo by di-
gefting it upon crude filings of that Metal, we obtaine a deep green Liquor,
which may be made ufe of inftead of
the Limpid oyle, to make the Diftin-
ction of the Liquors more confpicuous.

Sixthly, And for the fame purpofe
we often ufe inftead of clear water, a
ftrong Decoction of Brazill, or Logg-
wood, or elfe Red Inck it felf. I fay,
a *ftrong Decoction*, becaufe unlefs the Liquor be fo deeply ting'd, as to appear
Opacous in the Glafs, when it comes
into the flender pipe, its Colour will
be fo diluted, as to be fcarce difcernable.

Seventhly,

Seventhly, In the shape of the Glass
Vessel, we need not be Curious; though
that of a wide Mouth'd Jarr, exprefs'd
in the Scheme, be for some uses more
convenient than other shapes. The
depth of these Glasses, and the length
of the Pipes must be determin'd by the
Experiments, about which one means
to imploy them. To make out the
first Paradox already prov'd, a Glass of
about five or six Inches deep, and a
Pipe about as many Inches long, will
serve the turn: but for some others of
the following Experiments, tall Cylindrical Glasses will be requisite ; and
for some, Broad ones likewise will be
Expedient.

Eighthly, One must not be discou-
rag'd by not being able at the first
or second time, to suck up oyle of
Turpentine to the due height, and stop
it with ones finger from relapsing ; but
one must try again, and again ; especi-

ally

ally fince many Tryals of this kind may
be made in a few Minutes : and for
Beginners 'tis a fafe and good, though
not the fhorteft way, to fuck up rather
more Liquor then one judges will be
needful ; becaufe having fill'd the Pipe
to that height, you may by letting in
the Aire warily and flowly, between
the Orifice of the Glafs and the pulp
of your finger, fuffer fo much Liquor
to run out of the Pipe, as will reduce
it to the height you defire ; and there,
by clofe ftopping the Orifice with your
finger, you may keep it fufpended as
long as you pleafe, and immerfe it into
any Heterogeneous Liquor, and take
it out again at pleafure without fpilling
any of it.   By which flight Expedient
alone, I can decline feveral Difficulties,
and do many things, which, according
to *Monfieur Pafchal's* way, require a
great deal of Trouble and Apparatus
to be perform'd.

Laſtly, In ſuch Experiments where
it may be of uſe, That there be a conſiderable diſparity betwixt the two un-
mingled Liquors, we may (as is above
intimated) inſtead of fair water, imploy
Oleum Tartari per deliquium, and tinge
it with Brazill or Chochineele; from
either of which, but eſpecially from
the latter, it will obtaine an exceeding
deep Redneſs: and where one would
avoid ſtrong ſents and oylineſs, he may,
if he will be at the Charge, imploy
oyle of Tartar *per deliquium,* inſtead of
fair water, and highly Rectified Spirit
of Wine, inſtead of oyle of Turpentine.
For theſe two Liquors, though they will
both readily mingle with water, will
not with one another; and if a great
quantity of ſome other Liquor be to
be ſubſtituted for ſimple water, when
theſe Chymical Liquors are not to be
had in plenty, one may imploy (as we
have done) a very ſtrong Solution made
of

of Sea-falt, and filtred through Cap-
paper : · this Brine being near about as
Limpid as common water, and farre
heavier than it. And for a Curiofity,
we have added to the two lately mentioned Liquors (oyle of Tartar, and
Spirit of Wine)fome oyl of Turpentine,
and thereby had three Liquors of different Gravities,which will not by fhak-
ing,be brought fo to *mingle*,as not quickly to part again,& retire each within its
own Surface ; and by thrufting a Pipe
with water in the bottom of it (placing
alfo ones finger upon the upper Orifice)
beneath the Surface of the lowermoft
of thefe Liquors, and by opportunely
raifing or depreffing it, one may fomewhat vary the Experiment in a way not
unpleafant , but explicable upon the
fame grounds with the reft of the Phænomena mentioned in this Difcourfe.

PARADOX. II.

That a lighter Fluid may gavitate or weigh upon a heavier.

I Know that this is contrary to the com‧ mon opinion, not only of the Schools, but ev'n of divers hodiern Mathematici‧ ans, and Writers of Hydroſtaticks; ſome of whom have abſolutely rejeĉted this Paradox, though they do but doubt of the truth of the former.

But when I conſider, that whether the cauſe of Gravity be the pulſion of any ſuperior ſubſtance, or the Magneti‑ cal attraĉtion of the Earth, or what‑ ever elſe it be, there is in all heavy Bo‑ dies, as ſuch, a conſtant tendency to‑ wards the Centre, or lowermoſt parts of the Earth; I do not ſee why that
tendency

tendency or endeavour fhould be deftroy'd by the interpofition of any o-
ther heavy Body; Though what would
otherwife be the effect of that endeavour, namely an approach towards
the Centre, may be hindred by another
Body, which being heavier then it,
obtains by its greater gravity a lower
place; but then the lighter Body tending downwards, muft needs prefs upon the heavier that ftands in its way,
and muft together with that heavier
prefs upon whatever Body it is that
fupports them both, with a weight
confifting of the united gravities of the
more, and the lefs heavy Body.

But that which keeps Learned Men
from acknowledging this Truth, feems
to be this, That a lighter Liquor (or o-
ther Body) being environ'd with a
heavyer, will not fall down but emerge
to the Top ; whence they conclude,
that, in fuch Cafes, it is not to be confidered

fidered as a heavy, but as a Light Body.

But to this I anfwer, That though in Refpect of the heavier Liquor, the lefs heavy may in fome fence be faid to be light; yet, notwithftanding that relative or Comparative Levity, it retains all its abfolute Gravity, tending downwards as ftrongly as before; though by a contrary and more potent Endeavour upwards of the contiguous liquor (whofe lower parts, if lefs refifted, are preffed upwards by the higher elfewhere incumbent; according to the Doctrine partly delivered already, and partly to be cleared by the proof of the next propofition,) its endeavor downward is fo furmounted that it is forcibly carry'd up. Thus when a piece of fome light wood being held under water, is let go and fuffer'd to emerge, though it he buoy'd up by the water, whofe fpecifick Gravity is greater, yet ev'n whilft it afcends it remains a heavy
Body;

Body ; fo that the aggregate of the water & the afcending wood weighs more
then the water alone would doe; And
when it floats upon the upper part of
the water, as part of it is extant above
the furface, fo part of it is immerſt
beneath it, which confirms what we
were faying, That a lighter Body may
gravitate upon a heavier.

And thus there is little doubt to
be made but that if a man ſtand in one
of the fcales of a Ballance with a heavy
ſtone ty'd to his hand, and hanging
freely by his fide, if then he lift that
weight as high above his head as he
can, notwithſtanding that the ſtones
motion upwards makes it feem a light
Body in refpeɕt of the Man whofe
Body it leaves beneath it, yet it dos
not, either during its afcent or after,
loofe any thing of its connatural weight.
For the Man that lifts it up ſhall feel
its tendency downwards to continue,
though

though his force, being greater than that tendency, be able, notwithftanding that tendency, to carry it up : and when it is aloft, it will fo prefs againft his hand, as to offend, if not alfo to bruife it ; and the Stone, and the Man that fupports it, will weigh no lefs in the Scale he ftands in, then if he did not at all fupport it, and they were both of them weigh'd apart.

Likewife, if you put into one Scale a wide mouth'd Glafs full of water, and a good quantity of pouder'd common Salt ; and into the other Scale, a Counterpoife to them both ; you may obferve, that, though at the beginning the Salt will manifeftly lie at the bottome, and afterwards by degrees be fo taken up into the Body of the Liquor, that not a grain will appear there ; yet neverthelefs (as far as I can judge by my Experiments) the weight in that Scale will not be diminifhed by the

weight

weight of as much Salt as is inceffantly either carried up, or fupported by
the reftlefs motion of the diffolving
Corpufcles of the water ; but both the
one and the other, (allowing for what
may evaporate) will concurrently gravitate upon the Scale that the glafs containing them leans on.

But of this more elfewhere. Now
to prove the propofition by the New
Method, we have propos'd to our felf
in this Difcourfe.

Take a flender Glafs pipe, and having fuck'd up into it fair water, to the
height of 3 or 4 Inches, ftop nimbly the
upper Orifice with your finger, and
immerfe the lower into a Glafs full of
oyle of Turpentine, till the Surface of
the oyle in the Veffel be fomewhat
higher than that of the water in the
Pipe ; then removing your finger,
though the Pipe do thereby become o-
pen at both Ends, the water will not

fall

fall down, being hinder'd by the preffure of the oyle of Turpentine. As will
be obvious to them that have attentively confider'd the Explication of the
former Paradox; there being but this
difference between this Experiment and
that there Explain'd, that here the water is in the Pipe, and the oyle in the
Veffel, whereas there the oyle was in
the Pipe, and the water in the Veffel.
And if you either poure more oyle into the Glafs, or thruft the Pipe deeper
into the oyle, you fhall foe that the water will be buoyed up towards the top
of the Pipe; that is, a heavier Liquor
will be lifted up by a lighter. And
fince, by the Explication of the firft
Propofition, it appears, that the Reafon why the Liquor is in this cafe rais'd
in the Pipe, is the Gravity of the Liquor that raifes it, we muft allow that
a lighter Liquor in *fpecie*, may by its
gravity prefs againft a heavier

E   And

And it agrees very well with our Explication, both of this, and of the firſt
Experiment; that as there, the Surface of the oyle in the pipe was always
higher than that of the water without
it, becauſe the oyle being the lighter Liquor, a greater height of it was requir'd
to make an *Æquilibrium*; ſo in our pre-
ſents Experiment, the Surface of the
Liquor in the Pipe will alwayes be
lower than that of the oyle without it.
ſee the ſecond For in the imaginary plain *
Figure. E F, the Cylinder of water
J G, contain'd in the Pipe *J* H, will,
by reaſon of its greater gravity, preſs
as much upon the part *J*, as the diſtill'd
oyle (*K E*, *J L*,) being a lighter Liquor, can do, upon the other parts of
the ſame ſuppos'd plain *E F*, though
the oyle reach'd to a greater height a-
bove it.

This ſecond Paradox, we have hitherto been diſcourſing of, may be alſo

ſo prov'd by what we formerly deliver'd, to make out the Truth of the
third part of the Lemma premiſed
to theſe Propoſitions.

But becauſe this and the former
Paradox are of importance, not only
in themſelves but to the reſt of this
Treatiſe, and are likely (in moſt Readers) to meet with indiſpoſition enough
to be receiv'd, I will ſubjoyn in this
place a couple of ſuch Experiments,
as will not, I hope, be unacceptable;
that I devis'd, the one to confirm this
ſecond Paradox, and the other to
prove the firſt.

Some of the Gentlemen now preſent may poſſibly remember, that a-
bout the end of the Year that preceded
the two laſt, I brought into this place
a certain new Inſtrument of Glaſs,
whereby I made it appear, that the
upper parts of water gravitate upon the
lower; which I did by ſinking a Body,

that

that was already under water, by pouring more water upon it.

But that Experiment belonging to other papers, I shall here substitute another perform'd by an Instrument, which though it makes not so fine a shew, may be more easily provided, and will as well as that other (though you were pleas'd to command that from me) serve to make out the same Truth; which I shall apply my self to do, as soon as I have, by an Improvement of the Expedient I am to propose, made good my late promise of confirming the second Paradox.

And before I can well draw an Argument from these Experiments, for either of the propositions to be prov'd by them, I must briefly repeat what I have elsewhere deliver'd already (on another occasion) touching the cause of the smoking

In certain Notes upon some of the Physico-mechanical Experiments, touching the Aire.

ing of such Bubbles. Namely that
the Bubble X. consisting
of Glass, which is heavier *Fig.* 3.
in *specie*, then Water ; and
Aire, which is lighter in *specie* then
Water ; and, if you please, also of
Water itself, which is of the same specifick Gravity with Water ; as long as
this whole aggregate of several Bodys
is lighter then an equal bulk of Water,
it will float ; but in case it grows heavier
then so much water, it must, according
to the known Laws of the Hydrostaticks, necessarily sinck, (being not o-
therwise supported.) Now when there
is any competent pressure (whether
produc'd by weight or otherwise,) upon the water, in which this Bubble is for
the most part, immers'd, because the
glass is a firm Body, & the water, though
a Liquor, either suffers no compression,
or but an inconsiderable one ; the Aire
included in the Bubble, being a springy
                                    and

and a very incompreſſible Body , will
be compell'd to ſhrink , and thereby poſſeſſing leſs Room, then it did
before, the contiguous water will ſuc-
ceed in its place ; which being a
body above a thouſand times heavier
then aire , the Bubble will thereby become heavier then an equall
Bulk of water , and conſequently
will ſink : but if that force or pref-
ſure be remov'd , the Impriſon'd Aire
will by its own Spring free it ſelf
from the intruding water ; and the
Aggregate of Bodys, that makes up the
Bubble, being thereby grown lighter
then an equal bulk of water, the ſubſided
bubble will preſently emerge to the
Top.

This Explication of the Cauſes of
the ſinking of Bubbles agrees, in ſome
things, with the Doctrine of the Learned Jeſuites *Kercher* & *Shottus*, and ſome
other writers, in the Acount they give
of

of thofe two Experiments that are
commonly known by the name, the
one of the Romane, the other of the
Florentine Experiments. But there are
alfo particulars wherein I (who have
never a recourfe to a *fuga Vacui*,) diffent from their Doctrine ; the principles I go upon, having invited and affifted me to make that Experiment, afford me fome new Phænomena, which
agree not with their Opinions, but do
with mine : but I forbear to mention
them here, becaufe they belong to o-
ther Papers ; and for the fame reafon I
omit fome acceffion of Ludicrous Phænomena (as they call them,) which I
remember I have fometimes added to
thofe, which our Induftrious Authors
have already deduc'd from thofe Experiments.

Thefe things being premis'd, I proceed to the confirmation of the fecond
Paradox, by the following Experiment.

E 4   Take

Take a long glass pipe, seal'd or otherwise exactly stop'd at one end and open at the other; (whose Orifice if it be no wider, then that it may be conveniently stop'd with a mans Thumb, the Tube will be the fitter to exhibit some other Phænomena.) Into this pipe pour such a quantity of common water, as that there may be a foot, or half a yard, or some other competent part left unfill'd, for the use to be by and by mention'd. Then having poiz'd a glass Bubble with a slender neck, in such a manner as that though it will keep at the Top of the water, yet a very little addition of weight will suffice to sinck it, put this Bubble thus poiz'd into the Tube; where it will swim in the upper part of the water, as long as it is let alone, but if you gently pour oyle of Turpentine upon it, (I say gently to avoid confounding the Liquors) you will perceive that, for a while, the

Bubble

Bubble will continue where it was: but
if you continue pouring on oyl, till it
have attain'd a sufficient height above
the water, (which 'twill be easie to
perceive, because those two liquors will
keep themselves distinct) d you shall
see the Bubble subside till it fall to the
Bottom, and continue there as long as
the oyl remains at the height above
the water.

The Reason of this Phænomenon,
according to our Doctrine, is this,
That the oyl of Turpentine, though a
lighter Liquor then water, yet gravitates upon the subjacent water; and by its
pressure forces some of it into the cavity
of the bubble at the open Orifice of its
neck, whereby the Buble, which was before but very little less heavy then an e-
qual Bulk of water, being by this accessi-
on made a little more heavy, must necel-
sarily sinck ; and the cause of its sub-
mersion, namely the pressure of the

oyle, continuing, it muſt remain at the bottom.

And to confirm this explication I ſhall add, that in caſe, by inclining the Tube or otherwiſe, you remove the Cylinder of oyl, or a competent part of it, (in caſe it were longer then was neceſſary,) the Bubble will again emerge to the Top of the water (for, as for the oyle, that is too light a Liquor to buoy it up;) which happens only be-cauſe the preſſure of the oyle upon the water being taken of, the Air, by vertue of its own ſpring, is able to recover its former Expanſion , and reduce the bubble to be as light as 'twas be-fore.

And now we may proceed to that other Experiment, by which we late-ly promis'd to confirm the firſt Para-dox. And in ſome regard this follow-ing Experiment has been preferr'd, as more ſtrange, to that I have been reci-ting.

ting. For it feem'd much lefs improbable, that of two Heterogeneous Liquors, the inferior fhould be prefs'd upon by the incumbent, which, though
lighter, kept in an intire body above it,
then that in water, which is a Homogeneous Liquor, and whofe parts mingle moft freely and exquifitely with one
another, the upper part fhould prefs
upon the lower ; and that they will do
fo, may appear by the Experiment it is
now time to fubjoyn.

Provide a long Tube and a poiz'd
Bubble, as in the former Experiment,
then having pour'd water into the
Tube, till it reach above 5 or 6 Inches
(for a determinate height is no way
neceffary) above the Bottome, caft
in the Bubble , which will not only
fwim , but if you thruft it down into
the water it will of it felf emerge to the
upper part of it. Wherefore take a
flender Wand, or a Wire, or a flender
glafs

glass pipe, or any such Body that is
long enough for your purpose, and with
it having thrust the bubble beneath the
Surface of the water, pour water slow-
ly into the Tube (whose Cavity will
not be near fill'd by the rod or wire) till
it have attain'd a competent height,
(which, in my last Tryals, was about
a Foot, or half a Yard above the bubble:) and you shall see, that the bubble,
which before endeavour'd to emerge,
will by the additional weight of the incumbent water, be depress'd to the bottom of the Tube. After which you
may safely remove the wire, or other
body that kept it from rising. For as
the weight of the Incumbent water
was that which made it sink, so that
weight continuing on it, the bubble
will continue at the bottom.

But yet it is not without cause, that
we imploy a wire, or some such thing,
in this Experiment, though we affirm

it

it to be onely the weight of the Incumbent water, that makes the Bubble finck.
For if you fhould pour water into the
Tube, to the height lately mention'd,
or ev'n to a greater, if you did not make
ufe of the Wire, it would not ferve the
turn; becaufe that as faft as you pour
in the water, the Bubble being left to
it felf, will rife together with it; and
fo, keeping always near the upper part
of the water, it will never fuffer the Liquor to be fo high above it, as it muft
be, before it can deprefs it. But to confirm, that 'tis the weight of the Superior water that fincks the Bubble, and
keeps it at the Bottom; if you take out
of the Tube a competent quantity of
that Liquor, and fo take of the preffure of it from the Bubble, this will
prefently, without any other help, begin to fwim, and regain the upper part
of the water, whence it may at pleafure be precipitated, by pouring back
into

into the Tube the water that was taken
out of it. And these Confirmations, added to the former Proofs of the first and
second Paradoxes, being we conceive
sufficient to satisfie Impartial Readers
of the Truth of them, we should presently advance to the next Proposition,
if we did not think fit to interpose here
a *Scholium.*

SCHOLIUM.

IT may perchance be wondred at,
why, since we lately mention'd our
having made some Tryals with oyle of
Tartar *per deliquium,* we did not in the
present Experiment, in stead of fair water, make use of that, it being a very
much heavier Liquor, and (though it
may be incorporated with express'd
oyles) unmingleable in such Tryals with
oyle of Turpentine. But to this I answer,

fwer, That ev'n in fuch flender pipes, as
thofe made ufe of about the firft Experiment, I found that oyle of Tartar was
ponderous enough to flow down, though
flowly, into the oyle of Turpentine at
one fide of the immers'd Orifice, whilft
the oyle pafs'd upwards by it along the
other fide of the pipe. And my know-
ledge of this could not but make me
a little wonder, That fo Curious a per-
fon, as *Monfieur Pafchall*, fhould fomewhere teach, That if a Tube of above
14 foot long, and having its Orifice pla-
ced 14 foot under water, be full of
Quickfilver, the fluid Metal will not
all run out at the Bottom of the pipe,
though the Top of it be left open to
the Aire, but will be ftop'd at a foot
high in the pipe. For the Impetus,
that its fall will give it, muft probably
make it flow quite out of the pipe: And,
not here to mention thofe Tryals of
ours with Quickfilver and flender
Tubes,

Tubes, that made me think this very
improbable, if we consider that the Experiment will not succeed with much
more favourable circumstances, betwixt
oyle of Turpentine and oyle of Tartar,
though the heavier of these two Liquors be many times lighter than quicksilver : It tempts me much to suspect,
that *Monsieur Paschall* never actually
made the Experiment, at least with a
Tube as big as his Scheam would make
one guess, but yet thought he might
safely set it down, it being very conse-
quent to those Principles, of whose
Truth he was fully persuaded. And
indeed, were it not for the impetus, the
Quicksilver would acquire in falling
from such a height, the Ratiocination
were no way unworthy of him.
But Experiments that are but spe-
culatively true, should be propos'd as
such, and may oftentimes fail in pra-
ctise; because there may intervene divers

very other things capable of making
them mifcarry; which are overlook'd
by the Speculator; that is wont to
compute only the confequences of that
particular thing which he principally
confiders ; As in this cafe our Author
feems not to have confider'd, that in
fuch Tubes, as the Torricellian Experiment is wont to be made in, the largnefs of them would make them unfit
for this Tryal.

And I have known Ingenious men,
that are very well exercis'd in making
fuch Experiments, complaine, that
they could never make this of Monfieur Pafchall's to fucceed. In which
attempts, that the fize of the Tubes
much contributed to the unfuccefful-
nefs of the Tryals, I fhall (without
repeating what has been already intimated to that purpofe) in the following part of this Difcourfe have
F oppor-

opportunity to manifeſt ; and withal, to adde as Illuſtrious a proof of this our beloved Paradox; as all that as yet we have yet given,

PARADOX

PARADOX III.

That if a Body contiguous to the water be altogether, or in part, lower than the highest level of the said water, the lower part of the Body will be pref's'd upward by the water that touches it beneath.

THis may be prov'd by what has been already delivered in the Explication of the firft Experiment.: For where ever we conceive the loweft part of the Body, which is either totally, or in part, immers'd in water, to be, there the imaginary Superficies being beneath the true Superficies, every part of that imaginary Superficies muft be pref's'd upwards, by vertue of the weight of the water incumbent on
all

all the other parts of the fame Superficies, and fo that part of it, on which
the immers'd Body chances to leane,
muft for the fame Reafon have an endeavour upwards. And if that Endeavour be ftronger then that wherewith the weight of the Body tends
downwards, then (fuppofing there be
no Accidental Impediment) the Body
will be buoy'd or lifted up. And though
the Body be heavier then fo much water, and confequently will fubfide, yet
that Endeavour upwards of the water,
that touches its lower part, is onely
rendred ineffectual to the raifing or
fupporting the body, but not deftroy-
ed; the force of the heavy Body being from time to time refifted, and re-
tarded by the water, as much as it
would be if that Body were put into
one Scale, and the weight of as much
water, as is equal to it in bulk, were
put into the other.

To

To confirm this, we may have re-course to what we faid in the Explicati-on of the fecond Experiment. *Fig.* 1,2. For in cafe the flender pipe, wherein the water is kept fufpended, be thruft deeper into the oyl,or in cafe there be more oyle pour'd into the Veffel; the water will be impell'd up higher into the pipe ; which it would not be, if the oyle, though bulk for bulk a light-er Body, did not prefs againft the low-er Surface of the vvater, (where, alone, the two Liquors are contiguous,) more forcibly then the water by its gravity tends dovvnvvards. And even vvhen the Liquors reft in an *Æquilibrium;* the oyle continually preffes upvvards, a-gainft the lower Surface of the water; fince in that continual endeavour up-vvards confifts its conftant refiftance to the continual endeavour that the gra-vity of the water gives it to defcend. And fince the fame Phænomenon hap-

pens,

pens, whether we fufpend water in
oyle, as in the fecond Experiment, or
oyle in water, as in the firft: it appears, that the propofition is as well
applicable to thofe cafes, where the fuftein'd Body is fpecifically heavier, as
to thofe where 'tis fpecifically lighter
then the fubjacent fluid.

But a further and clearer proof of
this Doctrine will appear in the Explication of the next propofition. In
the mean time, to confirm that part of
our Difcourfe, where we mention'd the
Refiftance made by the water to Bodies that finck in it, Let us fuppofe,
in the annexed Figure, That
the pipe *E F* contains an oyle *Fig.* 4.
fpecifically heavier then water;
(as are the oyls of Guaiacum, of Cin-
namon, or Cloves, and fome others,)
and then, That the oyle in the pipe,
and the water without, being at reft
in an *Æquilibrium,* the pipe be flowly
rais'd

rais'd towards the Top of the Veſſel.
'Tis evident, from our former Doctrine, and from Experience too, that
there will run out drops of oyle, which
will fall from the bottom of the pipe,
to that of the Veſſel; but far more
ſlowly then if they fell out of the ſame
pipe in the Aire.

Now to compute how much the
preſſure of the water againſt the lower
parts of the drop amounts too, let us
ſuppoſe the drop to be G, to whoſe
lowermoſt part there is contiguous, in
any aſſignable place where it falls, the
imaginary Superficies H J. 'Tis evident, That if the drop of oyle were
not there, its place would be ſupplied
by an equal bulk of water; which being of the ſame ſpecifick Gravity with
the reſt of the water in the Veſſel, the
Surface H J would be laden every
where alike; and conſequently no part
of it would be diſplac'd. But now,
the

the drop of oyle being heavier then so much water, that part of the imaginary superficies, on which that drop leans, has more weight upon it, then any other equal part of the same Superficies; and consequently, will give place to the descending drop. And since the case of every other suppos'd Surface, at which the drop can be conceiv'd to arrive in its descent, will be the same with that of the Superficies HJ; it will for the Reason newly given, continue falling till it comes to the bottom of the Vessel which will suffer it to fall no further. And in case the drop G were not, as we suppose it, of a substance heavier in *specie* then water, but just equal to it, the contiguous part of the Superficies HJ would be neither more nor less charged then the other parts of the same Supeficies; and the part lean'd on would be neither deprefs'd nor rais'd, but the drop G would

would continue in the same place. And so we may prove, (what is affirm'd by *Archimedes* , and other Hydroſtatical Writers) That a Body æquiponderant *in ſpecie* to water, will reſt in any aſſignable place of the water where 'tis put.

And (to proceed further) ſince, if the drop *G* were of a matter but æquiponderant to water it would not ſinck lower at all , no more then emerge ; it follows, that though being heavier *inſpecie* then water, it will fall , yet the gravity upon whoſe account it falls, is no more then that by which it ſurmounts an equal bulk of water ; (ſince, if it were not for that overplus, the reſiſtance of the water would hinder it from falling at all :) and conſequently, it looſes in the water juſt as much of the weight it would have in the aire, as ſo much water, weigh'd likewiſe in the ſame aire, would amount to.

Which is a Phyſicall Account of'

that

that grand Theorem of the Hydrostaticks, which I do not remember that I
have seen made out in any Printed
Book, both solidly and clearly; The
Learned *Stevinus* himself, to whom the
later Writers are wont to refer, having but an obscure (and not Physical)
demonstration of it.

And, because this Theorem is not
only very noble, but (as we elsewhere
manifest) very useful, 'twill not be a-
miss to add, That it may easily be confirm'd by Experiment.

For if you take (for instance) a piece
of Lead, and hang it by a Horse haire
(that being suppos'd very near æquiponderant to water) from one of the
Scales of an exact Ballance; and,
when you have put a just Counterpoize
in the other Scale, suffer the Lead to
sinck in a vessel of water, till it be perfectly covered with it, but hangs freely in it, the counterpoize will very
much

much preponderate. And, part of the
Counterpoize being taken out till the
Ballance be again reduc'd to an *Æqui-*
librium, you may eafily (by fubducting
what you have taken out , and comparing it with the whole weight of the
Lead in the aire) find what part of its
weight it loofes in the water. And then
if you weigh any other piece of the
fame Lead, fuppofe a Lump of 12
ounces, and hang it by a Horfe haire at
one fcale, you may be fure that by put-
ing into the other fcale a weight lefs
by a twelfth part, (fuppofing Lead to
water to be as twelve to one) that is
eleven *ounces,* though the weights be
farr from an *Æquilibrium* in the Aire,
they will be reduc'd to it when the
Lead is cover'd with water.

The preffure of water againft the
lower part of the Body immers'd in it
may be confirmed by adding ; That
we may thence deduce the caufe of
the

the emergency of wood and other Bodyes lighter then water; which though
a familiar Effect, I have not found its
caufe to have been fo much as enquired
into by many, nor perhaps to have been
well rendred by any. If we fuppofe
then that the pipe be almoft fill'd, not
with a fincking but a fwimming oyle, as
oyle of Turpentine, if, as in the firft Experiment, the lower orifice be thruft under water, (to a far lefs depth then that
of the oyle in the pipe) and the upper
be flovvly unftop'd, the oyl vvill (as vve
formerly declar'd) get out in drops at
the bottom of the pipe. But to determine vvhy thefe drops, being quite cover'd and furrounded vvith vvater, and
prefs'd by it as vvell dovvnvvards as up-
vvards, fhould rather emerge then defcend, I fhall not content my felf to fay,
that vvater *in fpecie* heavier then this
kind of oyle ; For, befides that in fome
cafes (e're long to be mention'd) I have
made

made the water to deprefs ev'n this
kind of oyle, and befides that 'tis not
every piece of wood lighter in *fpecie*
then water that will float upon water,
how fhallow foever it be : The Queftion is how this præpollent Gravity of
the water comes to raife up the oyle,
though there be perchance much more
water, for it to break its way thorough,
above it, then beneath it.

. The Reafon then of the emerfion of
Lighter Bodies in heavier fluids, feems
to be this, That the endeavour upwards of the water, contiguous to the
lower part of the Body, is ftronger then
the endeavour downwards of the fame
Body, and the water incumbent on it.
As, in the former Scheme, fuppofing
the Drop *G* to be the oyle of Turpentine, and to touch the two imaginary
and parallel plains *H J*, *K L* ; 'tis evident, that upon the lower part of the
Drop, *N*, there is a greater preffure of
water,

water, then upon the upper part of the same Drop, *M* : becaufe that upon all the furface *K L*, there is but an uniform preffure of water *A K B L*, and upon all the parts of the furface *H I*, there is a greater weight of water *A H B I*, except at the part *N* ; for there the oyle *G*, being not fo heavy as fo much water, the oyle being expos'd to a greater preffure from beneath, then its own gravity (and that of the water incumbent on it) will enable it to refift, muft neceffarily give way and be impell'd upwards. And the cafe being the fame between that and any other parallel plain, wherefoever we fuppofe it to be in its afcent, it muft confequently be impell'd further and further upwards till it arrive at the Top ; and there it will float upon the water. Or, (to Explicate the matter without Figures) when a fpecifically lighter Body is immers'd under water, it is prefs'd

againſt

againſt by two pillars of water; the
one bearing againſt the upper, and the
other againſt the lower part; and becauſe the lengths of both theſe Pillars
muſt be computed from the Top of
the water, the lower part of the
immers'd boᴅy, muſt be preſs'd upon by a Pillar longer then the upper part by the thickneſs of the immers'd Body; and conſequently muſt
be preſs'd more upwards then downwards. And by how much the greater
diſparity of ſpecifick Gravity there is
betwixt the water and the emerging
Body, by ſo much the ſwifter (cæteris
paribus) it will aſcend: becauſe ſo much
the more will there be of preſſure
upon all the other parts of the imaginary ſurface, then upon that part that
happens to be contiguous to the Bottom of the aſcending Body.

And upon the ſame Grounds we may
give (what we have not yet met with)

a

a good folution of that Probleme, propos'd by Hydroftatical Writers, why,
if a Cylindrical ftick be cut in two
parts, the one as long again as the other,
and both of them, having been detain'd
under water at the fame depth, be let
go at the fame time and permitted to
emerge, the greater will rife fafter then
the lefler. For fuppofe one of thefe
Bodies, as $O\ P$, to be two foot high, and
the other, $Q\ R$, to be half fo much,
and that the lowermoft Surfaces of
both be in the fame imaginary plaine,
parallel to the uppermoft furface of
the water and three foot diftant from
it ; in this cafe there will be againft the
lower part of each of the wooden
Bodies a preflure, (from the laterally
fuperior water) equal to that upon all
the other parts of the Imaginary plain,
whereto thofe Bodies are contiguous ;
But whereas upon the upper furface of
the fhorter Body, $Q\ R$, there will lean

a

a pillar of water two foot high, the pillar of the fame Liquor that will lean
upon the Top of the taller Body, *P O*,
will be but one foot high ; as the attentive confiderer will eafily perceive. So
that the wooden Bodys being lighter *in
fpecie* then water, both of them will be
impell'd upwards; but that compoun-
ded pillar, (if I may fo call it,) which
confifts of one foot of wood and two
foot of water, will by its gravity more
refift the being rais'd, then that which
confifts of two foot of wood and but
one foot of water : fo that the caufe of
the unequal celerity in the Afcenfion of
thefe Bodys confifts chieflly, (for I
would neither overvalue nor exclude
Concomitant Caufes) that the difference of the preffure againft the upper
and lower part of each body refpective-
ly is greater in one then in the other.

And hence we may probably deduce
a reafon of what we often obferve in the

G   Diftil-

Diſtillation of the oyles of Anniſſeeds, Cloves, and diverſe Aromatick vegetables, in Lembecks by the intervention of water ; for oftentimes, when the fire has not been well regulated , there will come over, beſides the floating Oyle, a whitiſh water, which will not in a long time become cleare. And as we have elſwhere taught, That whiteneſs to proceed from the numerous reflections from the oyly ſubſtance of the Concrete, by the heat of the fire broken into innumerable little Globuls, and diſpers'd through the Body of the water ; ſo the reaſon why this whiteneſs continues ſo long, ſeems to be chiefly (for I mention not ſuch things, as, the great ſurfaces that theſe little Globuls have in reſpect of their Bulk) that, becauſe of the exceeding minuteneſs of theſe Drops, the height of the water that preſſes upon the upper part, is almoſt equal to that of the water that

preſſes

preſſes againſt the lower part ; ſo that
the difference between theſe two preſſures being inconſiderable, it has power
to raiſe the Drops but very ſlowly,(in-
ſomuch that upon this ground I devis'd
a Menſtruum,wherewith I could mingle
oyle in drops ſo exceedingly minute,
that, ev'n when there was but a few
ſpoonfuls of the mixture, it would continue whitiſh for divers whole days together) though at length they will e-
merge ; and the ſooner, becauſe whilſt
they ſwim up and down , as they frequently chance to meet and run into
one another , they compoſe greater
Drops ; which are (for the Reaſon already given) leſs ſlowly impell'd up by
the water : at the Top of which, the
Chymiſt, after a due time, is wont to
find new oyl floating. But whether this
be any way applicable to the ſwimming
of the inſenſible particles of corroded
metals in *Aquafortis*, and other ſaline

Men-

Menſtruums, I muſt not now ſtay to enquire.

One thing more there is, that I would point at before I diſmiſs this Paradox; Namely, that, for the ſame Reaſon we have all this while deduc'd, when the emergent drop, or any other Body, floats upon the Top of the water, it will ſinck juſt ſo far, (& no farther) till the immers'd part of the float-
Fig. 5. ing Body be equal in Bulke to as much water as is e- qual in weight to the whole Body. For ſuppoſe, in the annexed figure, *Υ* to be a Cube of wood three foot high and ʻſix pound in weight ; this wood, being much heavier then Aire, will ſinck into the water, till it come to an imaginary ſuperficies, *X W*, where, having the poſition newly deſcrib'd, it will neceſſa- rily acquieſce. For all the other equal parts of the Superficies, *X, W,Ջ,* be- ing lean'd upon by pillars of water e-

qual in height to the part XA, or
WB, if the whole weight of the wooden
Cube be greater then that of as much
water as is equal to the immers'd part,
it muſt neceſlarily ſinck lower, becauſe
the ſubjacent part of the Surface (at
V,)will be more charg'd then any of the
Reſt. And, on the other ſide, if the
Cube were lighter then as much water
as that whoſe place the immers'd part
takes up ; it muſt by the greater preſſure of the water upon the other parts
of the imaginary ſuperficies XW, then
upon that contiguous to the wood, (as
at V)be impell'd upward,til the preſſure
of the whole wood upon the part it
leans on,be of the ſame degree with that
of the reſt of the water,upon the reſt of
the ſuperficies: and conſequently be the
ſame with the water, whoſe place the
immers'd part of it takes up. The
lightneſs of that immers'd part, in
reſpect of ſo much water, being re-

compenc'd

compenc'd by the weight of the unimmers'd part , which is extant a-
bove the Superficies of the water.
And we fee, that when a piece of wood
fals into water, though, by the impetus it acquires in falling, it paffes
through divers imaginary plains that
lye beneath its due ftation ; yet the
greater preffure, to which each of
thofe plains is expos'd in all its other
parts, then in that which is contiguous
to the Bottom of the wood, dos quickly impel it up again, till, after fome e-
merfions and fubfidings, it refts at length
in fuch a pofition, as the newly explicated Hydroftatical Theorem affignes
it.

SCHOLIUM.

S C H O L I U M.

THis Ingenious Propofition (about floating Bodys) is taught and prov'd after the manner of *Mathematicians*, by the moft fubtle *Archimedes* and his Commentators : and we have newly been endeavouring to manifeft the Phyfical reafon why it muft be true. But *partly* becaufe the Propofition ought to hold, not only in fuch intire and homogeneous Bodyes as men exemplifie it in, (fuch as a piece of wood, or a Lump of wax) but in all Bodyes, though of a concave figure, and made up of many Bodys of never fo differing natures ; (and perhaps fome of them joyn'd together only by their fuperincumbency upon one another) and partly becaufe that a Truth, which is one of the main and ufefulleft of the Hydroftaticks, and may be of fo much

impor-

importance to Navigation, has no-
yet (that I know of) been attemt
pted to be demonſtrated otherwiſe then
upon Paper : it will not be amiſs,
for the ſatisfaction of ſuch of thoſe
whom it may concern,as are not vers'd
in Mathematical *Demonſtrations*, to add
an Experiment which I made to prove
it *Mechanically* ; as exactly as is neceſſary for the ſatisfaction of ſuch perſons.

After (then) having imploy'd ſeveral
Veſſels,ſome of wood, ſome of Laton,
and ſome of other materials, to compaſs what I deſir'd ; we found glaſſes
to be the moſt commodious we could
procure. And therefore filling a large
and deep glaſs to a convenient height
with fair water , we plac'd in it another deeper glaſs,ſhap'd like a *Goblet* or
Tumbler, that it might be the fitter for
ſwimming; and having furniſh'd it firſt
with *Ballaſt*, and then, for merryment
ſake,

fake, with a wooden Deck, by which a
tall Maft, with a Sayle faften'd to it,
was kept upright ; we fraughted with
wood, and by degrees pour'd Sand into
it, till we had made it finck juft to the
Tops of certaine confpicuous marks,
that we had faften'd on the outfide of
the Glafs to oppofite parts thereof.
Then obferving how high the water
reach'd in the larger Glafs, (which by
reafon of the Veflels Tranfparency
was eafie to be feen) we carefully
plac'd two or three markes in the
fame level with the Horizontal Surface of the water ; and taking out
the floating Veflel , as it was, with
all that belong'd to it, and wiping the
outfide dry, we put it into a good paire
of fcales, and having found what it a-
mounted to, we weigh'd in a competently large Viol (firft counterpoiz'd
apart) fo much water , (to a graine, or
thereabouts,) and pouring this water
into

into the large Glaſs above mentioned,
we found it to reach to the marks that
we had faſtened to the outſide of the
Glaſs, and conſequently to reach to the
ſame height to which the weight of the
floating glaſs, and all that was added to
make it reſemble a Ship, had made it a-
riſe to. By which Experiment (wch we
tried, as to the eſſential parts of it, with
Veſſels of differing ſizes, ſhapes; and
ladings too, as Wood, Stone, Quickſilver, &c.) it appears, that the floating Veſſel it ſelf, with all that was in it,
or ſupported by it, was equal in weight
to as much water as was equal in
bulk to that part of the Veſſel which
was under water, ſuppos'd to be cut off
from the extant part of the ſame veſſel
by a plain continuing the Horizontal
Surface of the water: ſince the weight
of the floating Veſſel, which rais'd up
the water in the larger Veſſel to the
greateſt height it attain'd, was the ſame
with

with the weight of the water, which being pour'd into the larger veffel (when the other was taken out) rais'd the water therein to the fame height. We may alfo obtaine the fame end, by a fomewhat differing way, (which is the beft way in cafe the Veffels be too great; *viz.* to obferve, firft, by pouring in water out of a Bowle or Paile, or other Veffel of known capacity, as often as is neceffary to fill the great Veffel, or Ciftern, or Pond, to the Top, (or to any determinate height requir'd) and, next, letting out, or otherwife removing all that water, to put in its place the Veffel, whofe weight is to be found out. Thirdly, to let, or poure in, water till the Veffel be afloat, and by its weight raife the External water to the height it had before: And laftly, to examine how much this water, that was laft pour'd in, falls fhort in weight

of

of the water that was in it at firft, and
afterwards remov'd. For this difference will give us the weight of as much
water, as is æquiponderate to the whole
floating Veffel, whither fmall or great,
with all that it either carries or fufteins.
The Hydroftatical Theorem we have
been confidering, and the Experiments
whereby we have endeavour'd to confirm, or illuftrate it, may (*Mutatis mu-
tandis*) be apply'd to a Ship with all
her Ballaft, Lading, Guns, and Company; it holding generally true, *That*
(to exprefs the fence of the Propofition
more briefly) *the weight of a floating
Body, is equal to as much water, as its
immers'd part takes up the room of.* Whence
we might draw fome Arguments in favour of the Learned *Stevinus*, (for whofe
fake it partly was that I annexed this
Scholium) who, if I mif-remember not,
does fomewhere deduce as a Corollary
from

from certain Hydroſtatical *See* PARADOX
Propoſitions, That a whole *the ſixth.*
Ship, and all that belongs to it, and
leans upon it, preſſes no more nor leſs
upon the Bottome it ſwims over, then
as much water, as is equal in bulk to
that part of the Ship which is beneath
the Surface of the water.

PARA-

PARADOX IV.

That in the Afcenfion of water in Pumps, &c. there needs nothing to raife the Water, but a competent weight of an External Fluid.

THis Propofition may be eafily e-
nough deduc'd from the already
mention'd Experiments.   But yet, for
further illuftration and proof, vve vvill
add that vvhich follovvs.

Take a flender Glafs-pipe, (fuch as
vvas us'd about the firft Experiment)
and fuck into it about the height of an
Inch of deeply tincted vvater; and, nimbly ftopping the upper Orifice, immerfe the lovver part of the pipe into
a Glafs half fill'd vvith fuch tincted
vvater, till the Surface of the Liquor

in

in the pipe be an inch (or as low as you would have it) beneath that of the External water. Then pouring on oyle of Turpentine till it swim 3 or 4 Inches, or as high as you please above the vvater; loosen gently your finger from the upper Orifice of the pipe, to give the inclosed Aire a little intercourse vvith the External, and you shall see the tincted vvater in the pipe, to be impell'd up, not only higher then the Surface of the External vvater, but almost as high as that of the External oyl, through vvhich (it being transparent and colourless) the Red Liquor may be easily discern'd.

Novv in this case it can't be pretended, That the ascent of the water in the pipe proceeds from Natures abhorrency of a *Vacuum*; since the pipe being full of aire, and its Orifice untopp'd, though the vvater should not ascend, no danger of a *Vacuum* vvould ensue; the aire and the vvater remaining contiguous as before. The

The true Reaſon then of the aſcent of the water, in our caſe, is but this, That upon all the other parts of the Imaginary Superficies, that paſſes by the immers'd Orifice of the pipe, there is a preſſure partly of water, and partly of the oyle ſwimming upon that water, amounting to the preſſure of 4 or 5 inches of water ; whereas upon that part of the ſame ſuperficies whereon the Liquor contain'd in the pipe leans, there is but the preſſure of one inch of water, ſo that the parts near the immers'd Orifice muſt neceſſarily be thruſt out of place by the other parts of water that are more preſs'd ; till ſo much Liquor be impell'd up into the pipe as makes the preſſure on that part of the Imaginary Superficies, as great as that of the oyle and water on any other equal part of it : and then, by Vertue of the *Æquilibrium,*(often mention'd) the water will riſe no further ; and, by vertue of the

ſame

same *Æquilibrium*, it will reſt a little beneath the Surface of the External oyle,
becauſe this laſt nam'd Liquor is leſſe
heavy, bulk for bulk, then water.

And by this we may be aſſiſted to
give a reaſon of the Aſcenſion of water
in ordinary ſucking Pumps. For as the
oyle of Turpentine, though a lighter
Liquor then water, and not mingleable
with it, does, by leaning upon the Surface of the External water, preſs up the
water within the pipe, to a far greater height then that of the External water it ſelf: ſo the Aire, which, though
a far lighter Liquor then oyle of Turpentine, reaches I know not how many
Miles high, leaning upon the Surface
of the water in a Well, would preſs it
up into the Cylindrical Cavity of the
Pump, much higher then the External
water it ſelf reaches in the Well, if it
were not hinder'd.

Now that which hinders it in the

H   Pump,

Pump, is either the Sucker, which
fences the water in the Pump from the
preffure of the External aire, or that
preffure it felf. And therefore, all
that the drawing up of the Sucker needs
to do, is, to free the water in the Pipe
from the impediment to its Afcent,
which was given it by the Suckers leaning on it, or the pillar of the Atmo-
fphæres being incumbent on it; as in our
Experiment, the fides of the pipe do
fufficiently protect the water in the
pipe from any preffure of the External oyle, that may oppofe its afcent.

And laftly, as the water in our
pipe was impell'd up fo high, and no
higher, that the Cylinder of water in
the pipe was juft able to ballance the
preffure of the water and oyle without
the pipe; fo in Pumps, the water does
rife but to a certain height, as about
33 or 34 foot: and though you pump
never fo long, it will be rais'd no
higher;

higher; becaufe at that height the preffure of the water in the Pump, upon
that part of the imaginary Superficies
that paffes by the lower Orifice of it,
is the fame with the preffure which
other parts of that imaginary fuperficies fuftaine from as much of the External water, and of the Atmofphære,
as come to lean upon it.

That there may be cafes wherein
water may be rais'd by fuction, not
upon the Account of the weight of the
aire, but of its fpring, I have elfewhere
fhovvn; and having likevvife in other
places, endeavour'd to explicate more
particularly the afcenfion of vvater in
Pumps; vvhat has been faid already
may fuffice to be faid in this place,
where 'tis fufficient for me to have
fhovvn, That vvhither or no the .A-
fcenfion of water *may have* other caufes,
yet in the .cafes propos'd, *it needs* no
more then the competent vvéight of

an

anExternalFluid,as is the Aire; vvhofe
not being devoid of gravity, the Co-
gency of our Experiments has brought
even our Adverfaries to grant us.

For confirmation of this, I will here
add, becaufe it now comes into my
mind, (what might perhaps be elfewhere fomewhat more properly mention'd) an Experiment that I did but
lightly glance at in the Explication of
the firft, and the *Scholium* of the fecond
Paradox.

In order to this I muft advertife,
That, whereas I there took notice, that
fome Ingenious men had complain'd,
that, contrary to the Experiment propos'd by *Monfieur Pafchall*, they were
not at all able to keep *Mercury* fufpended in Tubes, however very flender, though the lower end were deeply immers'd in water, if both their
ends were open: The Reafons of my
doubting, whether our Ingenious Author

thor had ever made or seen the Experiment, were, not only that it had been
unsuccesfully tryed, and seem'd to me
unlikely to succeed in Tubes more slender then his appear'd; but becaufe the
Impetus, which falling quickfilver gains
by the acceleration of motion it acquires
in its defcent, muft in all probability
be great enough to make it all run out
at the bottom of a Tube, open at both
ends, and fill'd with so ponderous a Liquor, though the Tube were very much
shorter then that propos'd by Monfieur
Pafchall.

This advertifement I premife to in-
timate, that, notwithstanding the hope-
leffnefs of the Experiment, as it had been
propos'd and tried, I might have reafon not to think it impoffible to perform, by another way, the main thing
defir'd; which was to keep Quickfilver fufpended in a Tube, open at both
ends, by the refiftance of the fubjacent

water.

water. For by the Expedient I am going to propose, I have been able to do it, even with a Liquor much lighter then water.

Finding then, that even a very short Cylinder of so ponderous a fluid, as *Mercury,* would, if it were once in falling, descend with an impetus not easy to be resisted by the subjacent Liquor, I thought upon the following Expedient to prevent this inconvenience. I took a slender pipe, the Diameter of whose Cavity was little above the sixth part of an Inch, and having suck'd in at the lower end of it somewhat lesse then half an inch of Quickfilver, and nimbly stopp'd the upper Orifice with my finger ; I thrust the Quickfilver into a deep glass of oyle of Turpentine, with a care not to unstop the upper Orifice, till the small Cylinder of quickfilver was 18 or 20 times its depth beneath the Surface of the oyle. For

by

by this means, when I unstopp'd the
pipe, the Quickſilver needed not (as
otherwiſe it would) begin to fall, as
having a longer Cylinder then was requiſite to make an *Æquilibrium* with
the other fluid. For by our Expedient the preſſure of the oyle was already
full as great, if not greater, againſt the
lower part of the Mercurial Cylinder,
as that which the weight of ſo ſhort a
Cylinder could exerciſe upon the contiguous and ſubjacent oyle. And accordingly, upon the removal of my finger, the Quickſilver did not run out,
but remain ſuſpended in the lower part
of the pipe. And as, if I rais'd it towards the Superficies of the oyle, the
Mercury would drop out for want of
its wonted Counterpoize; ſo, if I thruſt
the pipe deeper into the oyle, the increas'd preſſure of the oyle would proportionably impell up the Mercury towards the higher parts of the pipe,

which

which being again a little, and but a little, rais'd, the Quickfilver would fall down a little nearer the bottom of the pipe : and fo, with a not unpleafant fpectacle, the ponderous Body of quick-filver was made fometimes to rife, and fometimes to fall ; but ftill to float up-on the Surface of a Liquor, lighter then common Spirit of Wine it felf.

  But, befides that the Experiment, if the maker of it be not very careful, may eafily enough mifcarry, the diver-tifement it gives feldome proves laft-ing ; the oyle of Turpentine after a while infinuating it felf betwixt the fides of the pipe, and thofe of fo fhort a Cylinder of Mercury, and thereby difordering all. And therefore, though I here mention this Experiment, as I tryed it in oyle of Turpentine ; becaufe that is the Liquor I make ufe of all a-long thefe Paradoxes ; and becaufe alfo I would fhew that a lighter fluid then

water,

water, (and therefore why not aire, if its height be greatly enough increas'd?) may by its weight and preſſure, either keep the Mercury ſuſpended in pipes, or even raiſe it in them : Yet I found water (wherewith I fill'd tall glaſſes) a fitter Liquor then oyle for the Experiment ; in which though I ſought, and found ſome other Phænomena, yet becauſe they more properly belong to another place, I ſhall leave them unmention'd in this.

And ſince Experience ſhews us, that a Cylinder of Mercury, of about 30 Inches high, is æquiponderant to a Cylinder of water of about 33 or 34 foot high; its very eaſie to conclude, That the weight of the External aire, which is able to raiſe and keep ſuſpended 33 or 34 foot of water in a Pump, may do the like to 29 or 30 Inches of Quickſilver in the Torricellian Experiment.

PARA-

PARADOX V.

That the pressure of an External Fluid is able to keep an Heterogeneous Liquor suspended at the same height in several Pipes, though those Pipes be of very different Diameters.

THE contrary of this Proposition is so confidently asserted and believed, by those Mathematicians, and others, that favour the Doctrine of the Schools; That this perswasion of theirs seems to be the chief thing, that has hinderd men from acknowledging, that the Quicksilver in the Torricellian Experiment may be kept suspended by the Counterpoize of the external aire. And a famous writer, that has lately treated,

treated, as well of the Hydroſtaticks, as
of the Phænomena of the Torricellian
Experiment, dos rely ſo much upon the
falſehood of our Paradox, That, laying
aſide all other Arguments, he contents
himſelf to confute his Adverſaries with
one Demonſtration (as he calls it)groun-
ded on the quite contrary of what we
here aſſert. For his Objection runs
to this ſence. That if it were the
preſſure of the External Aire, that
kept the Quickſilver ſuſpended in the
newly mention'd experiment, the height
would not (as Experience ſhews it is)
be the ſame in all Cylindrical pipes,
though of very differing Bores. For,
ſuppoſing the height of the Mercurial
Cylinder, in a Tube of half an Inch
Diameter, to be 29 Inches; 'tis plain,
that a Mercurial Cylinder of the ſame
height, and three Inches in Diameter,
muſt weigh divers times as much as the
former; and therefore the preſſure

of

of the External aire, being but one and
the fame, if it be a juſt Counterpoize to
the greater Cylinder, it cannot be ſo to
the leſs.; and if it be able to keep the
one ſuſpended at 29Inches it muſt be a-
ble to keep the other ſuſpended at a far
greater height, which yet is contrary to
experience. And indeed this Objection
is ſo ſpecious, That, though I elſewhere
have already anſwer'd it, both by reaſon
and Experience, as far forth as it con-
cerns the Torricellian Experiment ;
Yet, to ſhew the miſtake on which it is
grounded, it may be very well worth
while to make out, our propoſed Paradox, (as that whoſe truth will ſufficiently diſprove that errour) by ſhewing
both *that* the Aſſertion is true , and
why it muſt be ſo.

Provide then a more then   *Fig.*6.
ordinarily wide mouth'd
Glaſs , cleer, and of a Convenient
depth ; into which having put a convenient

venient quantity of water, deeply
ting'd with Brazil or some other Pig-
ment, fit to the Orifice a broad but
thin Cork, in which, by burning or cut-
ting, make divers round holes of very
differing widenesses ; into each of
which you may thrust a glass Cylinder, open at both ends, and of a size fit
for the hole that is to receive it ; that
so the several pipes may be imbrac'd
by these several holes; And, as neare as
you can, make them parallel to one
another, and perpendicular to the superficies of the water, into which they
are to be immers'd. But we must not
forget, that, besides these holes, there
is an aperture to be made in the same
Corke (it matters not much of what figure or whereabouts) to receive the
slender end of a glass Funnel ; by
which oyl may be convey'd into the
vessel, when it is stopp'd with the Cork.
And in the slender part of this Funnel
we

we ufe to put fome Cotton-week, to break the violence of the oyl that is to be pour'd in, which might elfe diforder the Experiment. All this being thus provided, and the Cork (furnifh'd with its pipes) being fitted to the Orifice of the Veffel ; if at the Funnel you pour in oyl of Turpentine, and place the Glafs betwixt your eye and the Light ; you may, through that tranfparent Liquor, perceive the Tincted water, to be impell'd up into all the pipes, and to rife uniformly in them. And, when this tincted Liquor has attain'd to the height of two or three, or more Inches, above the lowermoft Surface of the External oyl ; if you remove the Funnel, (which yet you need not do, unlefs there be yet oyl in it,) you may plainly perceive the water to reach as high, in one of the fmaller pipes, as in another three or four times as great ; and yet the

water

water in the feveral pipes (as 'tis evident) is fuftain'd, at that height above the level of the other water, by the preffure or counterpoize of the external oyle; which therefore being lighter *in fpecie* then water, will have its Surface fomewhat higher without the pipes, then that of the Tincted water within them. And if by the Aperture, that receives the Funnel, you immerfe, almoft to the Bottom of the oyle, the fhorter leg of a flender glafs Syphon, at whofe longer Leg you procure by Suction the oyle to run out; you fhall perceive, That, according as the depth and preffure of the External fluid decreafes, fo the water in the pipe will fubfide; and that uniformly, as well in the leffer as in the greater pipes.

The Reafon of this is not difficult to be render'd, by the Doctrine already deliver'd. For fuppofe *E F* to be the Surface of the water, both within and

and without the pipes, before any oyle was poured on it: if we then suppose the oyle to be poured in through the Funnel, its lightness in respect of water, wherewith it will not mingle , will keep it from getting into the cavity of the pipes L, M, N ; and therefore spreading it self on the outside of them above, it must necessarily, by its gravity, press down the Superficies of the external water, and impell up that liquor into the cavities of the pipes. And if we suppose the pouring on of the oyl to be continued till the uppermost surface of the oyl be raised to G H, and that of the external water deprefs'd to I K, (or thereabouts,) an imaginary plain passing along the lower Orifices of the pipes ; I say , the tincted waters in the pipes ought to have their uppermost Surfaces in the same level , notwithstanding the great inequality of their Bores. For that part of the Surface

I K,

IK, which is comprehended within the Circular Orifice of the greateſt pipe *L*, is no more charged by the incumbent water, then any other part, equal to that Circle of the ſame Imaginary Superficies, is by the water or oyle incumbent on it ; (and conſequently, no more then the part comprehended within the circle of the ſmall pipe *N*, is by the water contain'd in that ſmall pipe ;) the external oyle having as much a greater height upon the Superficies *I K*, then the water within the pipe, as is requiſite to make the two Liquors Counter-ballance each other, notwithſtanding the difference of their ſpecifick Gravities. And though the pipe *L* were twice as bigg, it would Charge the ſubjacent plain *I K* no more, then the preſſure of the oyle on the other parts of the ſame imaginary Surface is able to reſiſt. And yet this preſſure of the External oyle ought not

I   to

to be able to raife the water in the flender pipe N, higher then the Surface Q in the fame Level with the Surface O. For, if the water were higher in the fmall pipe ; being a heavier Liquor then oyle, it muft prefs upon that part of the Surface I K, it leans on, with greater force then the external oyle upon the other parts of the fame plain I K ; and therefore with greater force then the weight of the External oyle could refift. And confequently, the water in the flender pipe muft fubfide, till its Surface be inferiour to that of the External oyle ; fince, till then, the difference of their fpecifick gravities cannot permit them to reft in an Æquilibrium. To be fhort ; It is all one, to the refiftance of the external oyl, how wide the Cylinder is that it fupports in the pipe ; provided the height of it be not greater in refpect of the height of the oyl, then the

the difference of the respective Gravities of those two Liquors requires.
For, so long the pressure of the Cylinder of water will be no greater, on that part of the Imaginary Superficies which it leans upon, then the pressure of the external oyle will be on all the other parts of the same Superficies; and consequently, neither the one, nor the other of those Liquors will subside, but they will both rest in an Æquilibrium.

But here it will not be amiss to note, First, that it is not necessary that the Glass Cylinders *L, M, N,* should be all of the same length; since, the lower Orifice being open, the water will rise to the same height within them, whether the parts immers'd under the water be exactly of the same length or no.

And Secondly, That throughout all this Discourse, and particularly in the

Expli-

Explication of this Paradox, we suppose, either that the slenderest pipes,
that are imploy'd about these Experiments, are of a moderate size, and not
exceeding small ; Or that, in case they
be very small, allowance be made in
such pipes for this property, That
water will rise in them to a greater
height, then can be attributed to the
bare Counterpoize of either the water
or the oyle, that impels it upwards and
keeps it suspended. But this difference
is of so little moment in our present In-
quiries, That we may safely neglect
it, (as hereafter we mean to do) now
we have taken this notice of it for
prevention of mistakes.

P A R A-

PARADOX VI.

If a Body be plac'd under water, with its uppermoſt Surface parallel to the Horizon; how much water ſoever there may be on this or that ſide above the Body, the direct preſſure ſuſtain'd by the Body (for we now conſider not the Lateral nor the recoyling preſſure, to which the Body may be expos'd if quite environ'd with water) is no more then that of a Columne of water, having the Horizontal ſuperficies of the Body for its Baſis, and the perpendicular depth of the water for its height.

And ſo likewiſe,

If the water that leans upon the Body be contain'd in pipes open at both ends; the preſſure of the water is to be eſtima-

ted

ted by the weight of a pillar of wa-
ter, whofe Bafis is equal to the lower
Orifice of the pipe,(which we fuppofe to
be parallel to the Horizon) and its
height equal to a perpendicular reaching
thence to the top of the water; though the
pipe be much inclin'd towards the Hori-
zon, or though it be irregularly fhap'd,
and much broader in fome parts, then
the faid Orifice.

S *Tevinus,* in the tenth Propofition of
his Hydroftatical Elements , having propos'd in more general termes,
the former part of our Paradox, anne-
xes to it a Demonftration to this purpofe.

Having firft fuppofed *A B C D,*
to be a folid Rectangular figure of
water, whofe *Bafis E F* is parallel to
the Horizon , and whofe height *G E* is
a perpendicular let fall from the uppermoft Surface of the water to the
lower-

lowermoſt ; His Demonſtration is this;

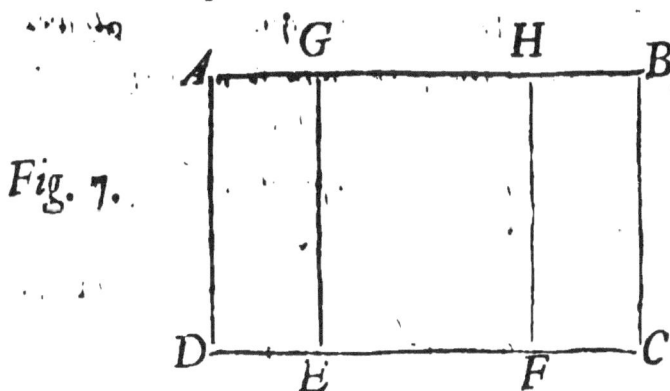

Fig. 7.

If the Bottom *E F* be charged with a greater weight then that of the water *G H F E,* that ſurpluſage muſt come from the adjoyning water ; there-fore; if it be poſſible, let it be from the water *A GE D,* & *H B C F;* which gran-ted, the Bottom *D E* will likewiſe have a greater weight incumbent on it, up-on the ſcore of the neigbouring water *G H F E,* then that of the water *A G E D.*   And , the reaſon being the ſame in all the three caſes, the Baſis *F C* muſt ſuſteine a grea-ter weight, then that of the wa-

I 4

ter

ter *H B C F.* And therefore the whole bottom *D.C* , will have a greater weight incumbent on it, then that of the whole water *A B. C D* ; which yet (*A B C D*, being a rectangular Body) would be abfurd. And by the fame way of reafoning you may evince, That the Bottom *E F.* fuftains no lefs a weight, then that of the water *G H F E.* And fo, fince it fuftains neither a greater weight, nor a lefs, it muft fuftein juft as much weight as the Columne of water *G H F E.*

This Demonftration of the Learned *Stevinus.* may well enough be admitted by a Naturalift (though, according to fome Hypothefes touching the Caufe and Nature of Gravity, it may faile of Mathematical exactnefs ;) and by it may be confirm'd the firft part of our propos'd Paradox. And fome things annexed by *Stevinus* to this Demonftration, may be alfo apply'd to

countenance

countenance the second. But beeause
this is one of the nobleft and usefulleft
Subjects of the Hydroftaticks, we
think it worth while to illuftrate, after
our manner, each of the two parts of
our Paradox by a fenfible Experiment.

Firft then, Take a flender Glafs-pipe
of an even Bore, turn'd up at one end like
the annexed Syphon. Into this
Syphon fuck oyl of Turpentine
till the Liquor have fill'd the fhorter
leg, and be rais'd 2 or 3 Inches in the longer. Then nimbly ftopping the upper
Orifice with your finger, thruft the
lower part of the Syphon fo farre into
a deep Glafs full of water, That the
Surface of the oyle in the longer leg of
the pipe, may be but a little higher then
that of the External water ; and, upon
the removal of your finger, you will
find the Surface of the oyle to vary but
little, or not at all, its former Station.

And'

And as, if you then thruſt the pipe a little deeper, you will ſee the oyle in the ſhorter leg to begin to be depreſs'd; ſo, if afterwards you gently raiſe the pipe toward the top of the water, you ſhall ſee the oyle not only regain its former ſtation, but flow out by degrees in drops that will emerge to the Top of the water. Now, ſince the water was able, at firſt, to keep the oyl, in the longer leg of the pipe, ſuſpended no higher, then it would have been kept by a Cylinder of water equal to the Orifice of the ſhorter leg of the pipe, and reaching directly thence to the Top of the water ; (as may be eaſily tried, by making a Syphon, where the ſhorter leg may be long enough to contain ſuch a Cylinder of water to counterpoize the oyl in the longer;) & ſince, when once, by the raiſing of the pipe, the height of the incumbent water was leſſen'd, the oyle did more then Counter-ballance

lance it ; (as appears by its flowing out of the Syphon;) we may well conclude ; That, though there were in the Veſſel a great deal of water, higher then the immers'd Orifice of the Syphon, (and it would be all one, though the Syphon were plac'd at the ſame depth in a pond or lake ;) yet, of all that water, no more did gravitate upon the Orifice, then that which was plac'd directly over it; which was ſuch a pillar of water, as the Paradox deſcribes.

And, by the way, we may hence learn ; That though water be not included in pipes, yet it may preſs. as regularly upon a ſubjacent Body, as if it were. And therefore we may well enough conceive a pillar of water, in the free water it ſelf, where there is nothing on any ſide, but the contiguous water, to bound the imaginary pillar.

But

But I had forgot to add, That the firſt part of our Paradox will hold, not only when the water, ſuperior to the Body it preſſes upon, is free ; but alſo, when it is included in Veſſels of never ſo (ſeemingly) diſadvantageous a ſhape. For, if you ſo frame the ſhorter leg of a Syphon, that it may expand it's ſelf into a funnel, like that of *Fig. 6.* employ'd about the proof of the foregoing (fifth) Paradox; (for which purpoſe the legs muſt be at a pretty diſtance from each other:) though you fill that Funnel with water, the oyle in the longer and ſlender leg of the Syphon will be able to reſiſt the preſſure of all the water, notwithſtanding the breadth of the upper part of the funnel. So that, ev'n in this caſe alſo, the Surface of the oyle in the longer leg, will be but a little higher then that of the water in the funnel.

For further Confirmation of this ;

we

we caus'd to be made a Syphon, so
shap'd, that one of the legs (which
were parallel, and of the same Bore,)
had in the midst of it a Sphære of
Glass, save that it communicated with
the upper and lower parts of the same
leg.

In the uniform leg of the Syphon,
we put a convenient quantity of oyle
of Turpentine, and into the other, as
much water as fill'd not only the lower part of it, but the Globular part too.
And yet we did not find, that all this
water was able to keep up the oyle in
the uniform leg, at a greater height
then if the leg that contain'd the water
had been uniform too ; as much of the
water in the Globe, as was not directly over the lower Orifice of it, being
supported by the lateral parts (if I may
so call them) of the same Globe. And,
if that leg were, instead of water, fill'd
with oyle, and the uniform leg with
water ;

water ; notwithftanding the far greater quantity of oyl, that was neceffary
to fill that leg, whereof the hollow
fphære was but a part ; the water in
the uniform leg would not be kept up,
fo much as to the fame height with the
oyle in the mifhapen leg.

But to make this matter yet the
more clear, we caus'd a Syphon to be
made of the Figure ex- *Figur.* 9.
prefs'd in the adjoyning
Scheme ; into which having poured a convenient quantity of Mercury,
till it reach'd in the fhorter leg *C D*,
almoft to the bottom of the Globulous
part *E*, and in the longer leg *A B*, to
an equal height: We afterwards,
poured a fufficient quantity of water
into the faid longer leg *A B*, which
drove away the Quickfilver, and impell'd it up in the fhorter leg till it
had half, or more then half, fill'd the
Cavity of the Globular part *E*; (which
                                    yet

yet we did not wholly fill with Quickfilver, becaufe the Tube *A B* was not
long enough for that purpofe;) and
then we obferv'd, that, notwithftanding the great weight of (that Body,
which is of all Bodies, fave one, the
moft Ponderous) Quickfilver, which
was contain'd in the lower part of the
fame leg of the Syphon, the furface of
the Quickfilver *H G*, was impell'd up
as high by the water in the Leg *A B*,
as the difparity of the fpecifick weights
of thofe two Liquors (whereof one is
about 14 times as heavy as the other)
did require: So that it appear'd not,
that, for all the great weight of
Quickfilver, contain'd in the Globulous Cavity *E*, there prefs'd any more
upon the flender and fubjacent part
E C of that leg, then as much as was
plac'd directly over the lower Orifice
of the faid Cavity *E*. So that the other,
and lateral parts of that Mercury, being

ing supported by the concave sides of the Glass, whereunto they were contiguous, the water in the leg *A B*, appear'd not any more press'd by the quickfilver, then if the leg *C D* had been, as well as the other, of an uniform bignefs ; and, by this means, if we had made the hollow Globe of a large Diameter, a small quantity of water, poured into the leg *A B*, might have been able to raife a quantity of quickfilver exceedingly much heavier then it felf. But then fo little water can raife the quickfilver, in fo broad a pipe, but to an inconfiderable height.

To make out the fecond part of our Paradox by an Experiment, we took three Glafs-pipes ; the one *Fig.* 10. made like a Bolt-head, with a round Ball and two oppofite Stemms ; the other was an irregular pipe, blown with an Elbow, wherewith it made an Angle ; and the third

was

was as irregularly fhap'd, as I could
get it blown; being in fome places
much broader, and in fome much nar-
rower then the lower Orifice of it.
And thefe two laft nam'd pipes had
thelr upper ends fo inferted into holes,
made fit for them in a broad piece of
Cork; that, when they were immers'd,
they made not right Angles, but very
oblique ones, with the Horizontal Surface of the Liquor. The other Glafs
likewife, which confifted of a great
Bubble, and two oppofite pipes, was
faftened to the fame Cork, which having before hand been made fit for a
wide mouth'd glafs of a good depth,
and half fill'd with water, was thruft
as a ftopple into the mouth of the faid
glafs, fo that the water afcended a pretty way into each of the three pipes by
their lower Orifices, which as well as
the upper we left open; Then a good
quantity of oyle of Turpentine being

K   pour'd

pour'd into the fame Veffel, through
a funnel, the water was by the incumbent oyle impell'd up to the height of
2 or 3 Inches in each of the three pipes.
Which argues, that, notwithftanding
their being fo unequal in bignefs, ånd
fo irregular in fhape, (infomuch that we
guefs'd one of them was 10 or 12 times
greater in one part, then in another,
or then it was even at the Orifice) the
water, contain'd in each of them, prefs'd
upon its lower Orifice *no more* (I do
not add, nor no lefs) then it would have
done if it had been a Cylinder, having
the Orifice for its Bafis, and the perpendicular depth of the water and oyle
above, for its height. For in cafe each
of the pipes had contein'd but fuch a
Cylinder of water, that water would
neverthelefs have had its uppermoft
Superficies at the fame height : and
on the other fide, it would have been
impell'd up beyond it, if its weight did
not

not as ſtrongly endeavour to depreſs the immediately ſubjacent water, as the preſſure of the External fluids endeavour'd to impel it up.

And ſince the height of the water was about the ſame in the ſeveral pipes, though two of them, being very much inclin'd, contain'd much more water then if they were erected : yet by the ſame way of reaſoning we may gather, That the imaginary plain, paſſing by the immers'd Orifice of either of theſe inclining pipes, ſuſtain'd no more of preſſure, then it would have done from a ſhorter Cylinder of water if erected. And indeed, in all theſe caſes, where a pipe either is broader in other places then at its lower Orifice, or inclin'd any way towards the Horizon, the weight of the contain'd Liquor is not all ſupported by the Liquor or the Body contiguous to the lower Orifice, but partly by the ſides of the pipe it ſelf. And

there-

therefore if, when in a slender pipe you have brought a parcel of oyl of Turpen-tine to be in an *Equilibrium* with the External water, as in the Experiment belonging to the first Paradox; If, I say, when this is done, you incline the pipe towards the sides of the Glass, You may indeed observe the Surface of the oyle in the pipe to be, as before, a little higher then that of the water without it: But you shall likewise see, That, though the Orifice of the pipe were not thrust deeper into the water, yet there will be a pretty deal of water got up into the pipe; because the oyle not leaning now upon the water only, as it did before, but partly upon the water, and partly upon the pipe, its pressure upon the subjacent water is considerably lessen'd; and there-by the external water, whose pressure is not diminish'd too, is able to impel up the oyl, and intrude for a little way into the pipe. But if you re·erect the pipe, the pressure of the oyle being then again ex-erted

erted upon the fubjacent water, it will be able to deprefs,and drive it again out of the Cavity of the pipe.

And to this agrees very well what we further try'd as follows : We caufed 3 pipes to be blown (fhap'd as the adjoyning Figures;) one having $Fig.$ 11. in it divers acute Angles ; the other being of a winding form, like a *fcrue* or *worm* of the Limbeck ; and the third very irregularly crooked;and yet each of thefe pipes having all its crooked parts, and fome of its ftreight & erected parts,fil'd with oyl of Turpentine; being thruft to a convenient depth under water and unftopp'd there, (after the manner already often declar'd) we found, that, according to our Paradox, the furface of the oyle in the pipe was higher than that of the water without it, as much as it would have been in cafe the pipe had been ftreight, (as we try'd by placing by the crookedeft of them a ftreight pipe with oyle in it)though the quantity

of

of the oyle, in one of these pipes, were perhaps three times as much as would have suffic'd, if the pipe had been strait: So that this surplusage of oyle did not press upon the subjacent water, (for if it had done so, the oyle would have run out of the pipe.) And I remember, that lifting up as much of one of these crook-ed pipes, as I thought fit, somewhat above the Surface of the water ; when the Superficies of the oyle in the pipe was not above half an inch higher then that of the water without it, I estimated that the crooked pillar of oyl, contain'd in that part of the *pipe* which was above the Surface of the water, was about 7 or 8 Inches long. So true it is, that the pressure of Liquors, contain'd in pipes, must be computed by the perpendicu-lar that measures their height, what ever be their length or bigness.

Scholium.

SCHOLIUM.

THE Learned *Stevinus*, having de-monstrated the Proposition we lately mention'd out of him, subjoyns divers confectaries of which the truth hath been thought more questionable, then that of the Theorem it self. And therefore he thought fit to add a kind of Appendix to make good a Paradox, which seems to amount to this. That If, in the Cover of a large Cylindrical Box, exactly closed, there be perpendicularly erected a Cylindrical Pipe open at both ends, and reaching to the Cavity of the Box; this Instrument being fill'd with water, the circular Basis of it will susteine a pressure, equal to that of the breadth of the Basis and height of the Pipe.

I

I chofe thus to exprefs this Theorem, [which might be, according to *Stevinus*, propos'd in more general terms,) becaufe this way of exprefling it will beft fuit with the fubfequent Experiment, and may confequently facilitate the ununderftanding of the Paradox. But though the Learned *Stevinus*'s aims were to be commended; who finding this Propofition doubted, feems to have had a great mind to give an Experimental Demonftration of it, and therefore propofes no lefs then five pragmatical Examples (as he calls them) to make out the truth of what he afferts; yet in this he hath been fomewhat unhappy, that that Experiment, which alone (for ought I can find) has been try'd of all the five, is rejected as incompetent, by thofe that profefs to have purpofely made tryal of it. And indeed, by reafon of the difficulty of bringing them to a practical examen,

I

I have somewhat doubted whether or
no this useful writer did ever make all
those Tryals himself ; rather then set
down the events, he suppos'd they must
needs have ; as presuming his conje-
ctures rightly deduc'd from a Demon-
strative Truth. Wherefore though another of the Experiments, he proposes,
be not free from difficulty, yet having,
by the help of an Expedient, made
it practicable, we are induc'd by its
plainness and clearness to prefer it to
what elfe he proposes to the same purpose.

　　We provided then a veffel of Laton,
of the figure exprefs'd in
the Scheme, and furnifh-　　*See Fig.*12.
ed it with a loofe Bottom
C D, made of a flat piece of wood cover'd with a foft Bladder and greas'd on
the lower fide neer the edges, that leaning on the rim of wood *GH*, contiguous
every where to the infide of the Laton,

　　　　　　　　　　　　　　it

it might be eafily lifted from off this
Rim ; and yet lye fo clofe upon it, that
the water fhould not be able to get out
between them: And to the midft of this
loofe bottom was faftned a long ftring,
of a good ftrength , for the ufe hereafter to be declared. The Inftrument
thus fitted, the water was poured in a-
pace at the Top *A B*, which , by its
weight prefling the falfe Bottom *C D* a-
gainft the fubjacent Rim *G H*, contributed to make the Veffel the more tight,
and to hinder its own pafling.   The
Veffel being fill'd with water we took
the forementioned ftring, one of whofe
ends was faftned to *I*, the middle part
of the loofe Bottom; and , tying the
other end *K* to the extremity of the
Beam of a good pair of Scales, we put
weights one after another into the oppofite fcale, till at length thofe weights
lifted up the falfe Bottom *C D* from
the Rim *G H*; and , confequently,
lifted

lifted up the Incumbent water ; which
prefently after ran down between them.
And having formerly, before we poured in any water , try'd what water
would fuffice to raife the Bottom *C D*,
when there was nothing but its own
proper weight that was to be furmounted ; we found , by deducting that
weight from the weight in the fcale ,
and comparing the Refidue with the
weight of as much water, as the cavity
of the broad , but very fhallow
Cylinder *B E C H G D F* would have
alone (if there had been no water in the
pipe *A I*) amounted to ; we found, I
fay, by comparing thefe ' particulars,
that the preflure upon *C D* was by fo
very great odds more, then could have
been attributed to the weight of fo
little water, as the Inftrument pipe and
all contain'd , in cafe the water had
been in an uniform Cylinder, and confequently a very fhallow one, of a Bafis

fis as large as that of our Inftrument,
That we could not but look upon the
fuccefs, as that, which though it did
not anfwer what the reading of *Stevi-*
nus might make a man expect; yet
may deferve to be further profecuted,
that whether or no the Paradox of *Ste-*
vinus (which not only fome others,
but the Learned Dr. *Wallis* himfelf
queftion) wil hold ; the Inquiry he has
ftarted, may be fo perfued, as to occafion fome improvement of this part of
Hydroftaticks : where, to define
things with certainty, will perhaps be
found a difficulter Task then at firft
glance one would think ; both becaufe
divers fpeculative things muft be taken
into confideration, whofe Theory has
not perhaps yet been clear'd, and becaufe of the difficulty that will be
found in practice by them that fhall go
about to make *Stevinus*'s Experiments,
or others of that fort with all requisite

fite Accuratenefs : As indeed, it is far eafier to propofe Experiments, which would in likelyhood prove what we intend, in cafe they could be made, then to propofe practicable Expedients how they may be made.

· PARA-

PARADOX VII.

*That a Body immers'd in a Fluid, fu-
ftains a lateral preffure from the Fluid ;
and that increas'd, as the depth of the
immers'd Body, beneath the Surface of
the Fluid, increafeth.*

THough I fhall not wonder if this
proposition feems ftrange enough
to moft Readers : yet I think I could
make it out by feveral wayes, and particularly by one that is plain and eafie,
being but that which follows.

 Take then a flender Glafs pipe
Fig. 13. (like that imployd about the
    firft Experiment;) and caufe
it to be bent within two or three Inches
of one end , fo that the longer and the
             fhorter

ſhorter legs, *E F* and *F G*, may make, as
near as can be, a right Angle at *F*; then
dipping the Orifice of the ſhorter leg
F G in oyle of Turpentine, ſuck into
the Syphon (if I may ſo call it) as
much of the Liquor, as will fill the
ſhorter leg, and reach two or three
Inches high in the longer ; then, nimbly
ſtopping the upper Orifice with your
finger, immerſe the lower part of the
Glaſs under water, in ſuch manner as
that the longer leg *EF* may make, as to
ſenſe, right Angles with (*A B*) the Horizontal Surface of the water, and the
ſhorter leg *F G* may be ſo far depreſs'd
under that Surface, That *I K*, the Superficies of the oyle in the longer leg,
be but a little higher then *A B*, that of
the external water. Then, removing
your finger, you may obſerve, That
the oyle in the Syphon will continue
(with little or no change) in its former
ſtation. By which it appears that there
is

is a lateral preffure of the water againft
the oyle contiguous to *G*, the Orifice of
the fhorter leg of the pipe,fince it is only that preffure that hinders the efflux
of the oyl at that Orifice,notwithftand-
ing the preffure of the perpendicular
Cylinder of oyle that would drive it
out.

And that this preffure of the perpendicular Cylinder doth really urge
the oyle in the fhorter leg to flow out;
you may learn by flowly lifting the
Syphon (without changing its former
pofture) towards the Surface of the
water. For as the lower leg comes
nearer and nearer to that Surface, (to
which, as I newly intimated, it is ftill
to be kept parallel) the oyle in the
Horizontal leg will be driven out in
drops, by the preffure of the other oyle
in the perpendicular leg.

That likewife before you begin to
raife the Syphon, the lateral preffure
of

of the water againſt the lower Orifice
of it is, at leaſt in ſuch Experiments,
near about the ſame with what would
be the perpendicular preſſure of a Cylinder of water, reaching from the ſame
Orifice G (or ſome part of it) to the
top of the water, may be gather'd from
hence, That the Surface of the oyle in
the longer leg will be a litle higher then
that of the external water, as (by reaſon
of the often mention'd comparative levity of the oyl) it would be, if we ſuppoſe,
That a pipe of Glaſs of the ſame bore,
and reaching to the top of the water,
being fitted to the Orifice of the Horizontal Leg (as in the annex'd figure the
Cylinder, G H) were fill'd with water.

And, to make out the latter part of
our propoſition, we need add no more,
then that, if you plunge the Syphon
deeper into the water, you ſhall find
the oyle , by the Lateral preſſure
of the water , driven by degrees

quite

quite out of the shorter leg into the longer : and if you thrust it yet deeper, you may observe that the longer leg will admit a Cylinder of water, upon which that of oyle will swim; the whole oyle alone being unable to counterballance the lateral pressure of the water at so great a depth.

By which last circumstance, it appears, that water has also a lateral pressure against water it self, and that increas'd according to its depth ; since otherwise the external water could not impel that in the Horizontal leg of the Syphon, into the perpendicular leg, though to doe so, it must surmount the weight or resistance of the whole cylinder of oyl, that must be hereby violently rais'd in the said perpendicular leg.

But if you gently raise the Syphon again, the lateral pressure of the water against the immers'd Orifice being diminish'd, (according as the distance

of

of that Orifice *G* from the Horizontal Surface, *A B,* comes to be leſſen'd,) the prevalent oyle will drive out the water, firſt out of the Longer leg, and then out of the ſhorter, and will at length flow out in drops at the immers'd Orifice, and thence emerge to the Top of the water.

Beſides, when the oyle in the Syphon does juſt counterballance the external water, if you keep the ſhorter leg parallel to the Surface of the water, and move the Orifice of it this way or that way, and place it nearer or further off from the middle or from the ſides of the Glaſs, (provided you keep it always at the ſame depth under the water,) you'l find the oyl in the longer Leg to continue (as to ſenſe) at the ſame height: Whence we may learn (what I have not yet found mention'd by any Writer,) That, ev'n in the mid'ſt of the water, we may ſuppoſe a

L 2 pillar

pillar of *water*, of a Basis equal to the side
of an immers'd Body, (and reaching to
the lowest part of it;) And that, though
this Imaginary aqueous pillar, such as in
our figure *G H*, be not included in any
solid Body or stable superficies ; neverthelefs its lower parts will have a lateral
preffure tending outwards, againſt the
imaginary fides , from the weight of
the water that is above thefe fubjacent
and lateral parts ; and will have that
preffure increas'd proportionably to the
height to which the imaginary pillar
reaches above them. Which obferva-
tion, being duely noted and apply'd,
may be of no mean ufe in the explication of divers Hydroſtatical phænomena.

And laſtly if, in ſtead of holding *E F*,
the longer leg of our Syphon, perpendicular, (and, confequently, the ſhorter parallel to the Horizon,). you va-
rioſly incline the former ; ſo as to
bring

bring it to make an obtufe or an Acute
Angle with the fuperficies of the water
A B; though by this means the fhorter
and immers'd leg, *F G*, will in Situa-
tion fometimes refpect the Bottom ,
and fometimes the top of the Glafs :
yet in all thefe oblique fituations of this
leg, and the immers'd Orifice of it,
G, the oblique preffure of the water
will fo much depend upon the height
of the Surface of the Liquor above the
Orifice, and fo much conform to the
obfervations already deliver'd, That
you fhall ftill fee the furface of the
oyle *I K*, in the longer pipe, to be a
little, and but a little fuperiour to that
of the external water, *A B*, and fo the
Æquilibrium betwixt the Liquor, or
Liquors, within the Syphon, and the
water without it, will ev'n in this cafe
alfo be maintain'd.

L 3   *Scholium*

S C H O L I U M.

REmembring on this Occafion an Experiment, which though it do not fhew what the precife quantity of Lateral preffure is, that the lower parts of the fluid may fuftain from the more elevated ; yet it may confirm the foregoing Paradox , and by its Phæ-nomena afford fome hints that may render it not unacceptable ; I fhall fub-joyn it, as I fet it down not long after I devis'd it.

In the firft place then, there was made a glafs Bubble with a flender neck ; and (in a word) of the figure exprefs'd in the annex'd *Fig.* 14.   Scheme ; This Bubble I caus'd to be fo poys'd, That, though it would float upon the water, yet the addition of a weight fmall enough would fuffice to make it finck.   **This**

This done, I provided a very large wide mouth'd Glafs, and caus'd to be fitted to it, as exactly as I could, a ftopple of Cork, which being ftrongly thruft in, would not eafily be lifted up. In the middle of this Cork there was burn'd, with a heated inftrument, a round hole; through which was thruft a long flender pipe of Glafs; fo that the lower end of it was a pretty way beneath the Cork, and the upper part of it was, as near as could be, at right Angles with the upper part of the faid Cork. And in an other part of the ftopple, near the edge, there was made another round hole, into which was likewife thruft another fmall pipe; whofe lower part reach'd alfo a pretty way beneath the Cork, but its upper part was but about two or three Inches high; and the Orifice of this upper part was carefully clos'd with a ftopple and Cement. Then the glafs veffel be-

ing

ing fill'd with water, and the pois'd
Bubble being made to float upon it, the
stopple or cover of the great glass vessel was put on, and made fast with a
close Cement, that nothing might get
in or out of the vessel, but at the long
slender pipe; which was fastned into
the Cork (as was also the shorter
pipe) not only by its own fitness to
the hole it pass'd through, but by a
sufficient Quantity of the same
Cement, carefully apply'd to stop all
crevesses.

The Instrument thus prepar'd, (and
inclin'd this or that way, till the floating Bubble was at a good distance from
that end of the long pipe, which rea-
ched a pretty way downwards beneath
the Surface of the water,) we began to
pour in some of that Liquor at the o-
pen Orifice of the pipe $E F$; and, the
mouth of the Vessel being exactly
stopp'd, the water for want of another
plac e

place to receive it, afcended into the
pipe through which it had fallen before. And, if I held my hand when
the water I had pour'd in was able to
reach but to a fmall height in the Cylinder, as for inftance, to the Superficies *J* ; the Bubble *X* would yet continue floating. But if I continued pouring till the water in the pipe had attain'd to a confiderable height above the
Surface of that in the Veffel, as if it
reach'd to K ; then the Bubble *X* would
prefently finck to the bottome of the
Veffel ; and there continue, as long as
as the water continued at fo great a
height in the pipe *E F*.

ThisExperiment will not only teach
us, That the upper parts of the water
gravitate upon thofe that are under
them, but (which is the thing we are
now to confirm) That in a Veffel, that
is full, all the lower parts are prefs'd
by the upper, though thefe lower be

not

not directly beneath the upper, but a-
side of them, and perhaps at a good distance from the Line in which they directly press: These things, I say, may
be made out by our Experiment. For
the Addition of the Cylinder of water
K J, in the pipe E F, makes the Bubble X subside; as the force or pressure
of any other heavy body upon the water in the vessel would do. And since
(as may be gather'd from the Reason
formerly given (in the Proof of the second Paradox) of the sincking of pois'd
Bubbles) the included aire in our Bubble was notably compress'd; it will
follow, that the Cylinder of water, KI,
did press the subjacent water in the
Vessel. For, without so doing, it could
not be able to compress the aire in the
Bubble. And since the said Bubble
did not swimme directly under or near
the pipe E F; but at one side of it, and
at a pretty distance from it, nay and
floated

floated above the lower Orifice, *F*, of
the pipe; 'tis evident that that Aqueous Cylinder, *J*K, does not only prefs
upon the water, or other Bodies that
are directly under it; but upon thofe
alfo that are laterally fituated in refpect
of it, provided they be inferior to it.

: And, according to this Doctrine, we
may conceive, that every affignable part
of the fides of the Veffel does fuftaine
a preffure, encreas'd by the encreafe of
that parts depth under water, and according to the largnefs of the faid part.
And therefore, if any part were fo weak,
as that it would be eafily beaten out or
broken by a weight equal to the Cylinder *I* K, (making always a due abate-
ment for the obliquity of the preffure)
it would not be fit to be a part of our
Veffel : Nay the Cork it felf, though
it be above the Surface of the water in
the Veffel; yet becaufe the water in
the pipe is higher then it, each of its
parts.

parts refifts a confiderable preffure proportionate to its particular bignefs, and
to the height of the water in the pipe.
And therefore, if the Cork be not well
ftopp'd in, it may be lifted up by the
preffure of the water in the pipe, if that
be fill'd to a good height. And if the
Cement be not good and clofe, the water will (not without noife) make it felf
a paffage through it. And if the ftopple G, of the fhorter pipe G H,(which
is plac'd there likewife to illuftrate the
prefent conjecture) do not firmly clofe
the Orifice of it, it may be forced out,
not without violence and noife. And,
for further fatisfaction,if,in ftead of the
ftopple G, you clofe the Orifice with
your finger, you fhall find it prefs'd upwards as ftrongly,as it would be prefs'd
downwards by the weight of a Cylinder of water of the breadth of the pipe,
and of a not inconfiderable height, (for
'tis not eafie to determine precifely,
                                              what

what height :) fo that (to be fhort) in
the fluid Body, we made our tryal with,
the preffure of the Superior parts was
communicated, not onely to thofe that
were plac'd *directly* under them, but
ev'n to thofe that were but *obliquely* fo,
and at a diftance from them.

I had forgot to confirm, that it was
the preffure of the fuperiour parts of
the water, that made our floating bubble finck, by fuch another circumftance
as I took notice of in fome of the former
Experiments; *viz.* that, when it lay
quietly at the bottom of the Veffel, if
by inclining the Inftrument we pour'd
off as much of the water in the pipe,
E F, as fuffic'd competently to diminifh
its height above the water in the Veffel
A B C D, the air in the bubble finding
its former preffure alleviated, would
prefently expand it felf, and make the
bubble emerge. And to fhow, That
the very oblique preffure.wch the bubble

ble suftain'd from the water in the pipe,
was not overmuch differing from that
which it would have fuftain'd from an
External force, or from the weight of
water plac'd directly over it; I caus'd
two fuch bubbles to be pois'd, and having put each of them into a long Cylindrical Glafs, open above, and fill'd
with water, upon which it floated, if we
thruft it down a little way it would (a-
greeably to what hath been *See the Proof of*
above related) afcend again; *the II. Paradox.*
fo that we were forc'd to thruft it down
to a good depth, before the preffure of
the incumbent water was great enough
to make it fubfide.

And perhaps it will not be impertinent to take notice, before we conclude, how the preffure of fuch differing fluids, as aire and water, may be
communicated to one another. For
having fometimes forborn to fill the
Veffel *A B C D* quite full of water, fo
that,

that, when the Cork was fitted to it, there remain'd in it a pretty quantity of aire, (as between the Surface $L M$, and the Cork) neverthelefs, if the ftopple or cork were very clofely put in, the preffure of the water that was afterwards poured into the pipe $E F$, from J to K, would make the bubble finck, little otherwife, for ought I took notice of, then if the Veffel had been perfectly fill'd with water; the aire (above $L M$,) that was both imprifoned and comprefs'd, communicating the preffure it receiv'd to the water contiguous to it.

PARA-

PARADOX VIII.

That water may be made as well to depress a Body lighter then it self, as to buoy it up.

HOw ftrange foever this may feem, to thofe that are prepoffefs'd with the vulgar Notions about gravity and levity : It need not be marvail'd at, by them that have confider'd what has been already deliver'd. For fince, in Fluid Bodies, the upper parts prefs upon the lower, and upon other bodies that lie beneath them. And fince, when a Body is unequally prefs'd by others, whether lighter or heavier then it felf, it muft neceffarily be thruft out of that place, where it is more

prefs'd

prefs'd, to that where 'tis lefs prefs'd ;
If that a parcel of oyle be by a contrivance fo expofed to the water, as that
the water prefles againft its upper Superficies, and not againft the undermoft
or lateral parts of it ; If we fuppofe
that there is nothing (whofe preflure is
not inferiour to that of the water) to
hinder its defcent, (fuppofing, withal,
that the oyle and water cannot pafs by
one another ; for which caufe, we make
ufe of a flender pipe ;) the oyle muft neceffarily give way downwards, and confequently be deprefs'd and not buoy'd
up. This is ealily exemplified by the
following Experiment.

Take a flender Glafs Syphon *E F*
G H, of the bore we have often
mention'd, whofe fhorter leg *Fig.* 15.
G H may be about 3 or 4 Inches long,
and as parallel as the Artificer can make
it to the longer *E F*; dip the fhorter leg
in oyle of Turpentine, till the oyle

M   quite

quite fill the shorter leg, and reach to
an equal height in the longer, as from *F*
to *J*. Then stopping the Orifice *E* of the
longer leg with your finger, and immer-
sing the replenish'd part of the Syphon
about an inch under water, you shall
perceive that as you thrust it lower and
lower, upon the removal of your finger,
the oyle in the shorter leg will be made
to sinck about an inch or somewhat
more ; and as afterwards you thrust
the pipe deeper, the oyle in the shorter
leg will, by the weight of the incumbent water, *H K*, be driven downward
more and more, till it come to the very bottom of the shorter leg ; whence,
by continuing the immersion, you may
impel it into the longer. The cause
of which Phænomenon, I suppose to
be already clearly enough assign'd, to
make it needless to add any thing here
about it.

It remains, that, before I proceed
to

to the next propofition, I add ; That,
to Exemplifie at once three Paradoxes,
(both this, and the next fore-
going, and the fecond) I caus'd   *Fig.* 16.
to be made a flender Glafs-pipe, of
the Figure exprefs'd in the annexed
Scheme, and having, by the lower
Orifice *L*, fuck'd into it as much oyle
of Turpentine, as reach'd in the lon-
geft leg, *N O*, as high as the Top of
the other part of the Glafs ; (namely,
to the part *P*, in the fame level with
the Orifice *L*,) I firft ftopp'd the upper Orifice of it, *O*, with my finger.
And then, thrufting it as before under water to a convenient depth, upon
the removal of my finger, the External water did firft drive away the oyle
that was in *L M*, that part of the
crooked pipe which was parallel to the
Horizon ; then it deprefs'd the fame
oyle to the bottom of the fhorter leg,
that is from *M* to *N* : And laftly,

it impell'd it all up into the longer
leg *N P O*, to what height I thought
fit. So that the oyle was prefs'd by
the water both laterally, downwards,
and upwards: the caufes of which are
eafily deducible from the Doctrine already deliver'd.

PARA-

PARADOX IX.

*That, what ever is said of positive Levity,
a parcel of oyle lighter then water, may
be kept in water without ascending
in it.*

TO make out what I have to re-
present about this Paradox the
more intelligible, the best way perhaps will be to set down the Considerations that induc'd me to judge the
thing it pretends to feasible. And in
order to this, it would be expedient to
consider, why it is that a Body lighter
in specie then water, being plac'd never so much beneath the Superficies of
that Liquor, will rather emerge to the

Top,

Top, then sinck to the bottom of it; if we had not already consider'd that problem in the Explication of the third Paradox. But being now allow'd to apply to our present purpose what hath been there deliver'd, I shall forth-with subjoyne, That 'twas easie e-nough for me to collect from hence, that, the Reason why it seems not pos-sible, That a parcel of oyle lighter then water, should without violence be kept from emerging to the Top of it, being this, *That since the Surface of a Vessel full of standing water is (Physically speak-ing) Horizontal, the water that presses a-gainst the lower part of the immers'd Body, must needs be deeper then that which pres-ses against the upper:* If I could so or-der the matter, that the water that leans upon the upper part of the Body should by being higher then the level of the rest of the water, have a height great enough to ballance that which presses
against

againſt the lower, (and the Bodies not ſhift places, by paſſing one by the other) the oyle might be kept ſuſpended be-twixt two parcels of water.

To reduce this to practiſe, I took the following courſe ; having ſuck'd into a ſlender pipe (ſuch as that im-ploy'd about the firſt experiment)about an Inch of water, and kept it ſuſpended there by ſtopping the Orifice of the pipe ; I thruſt the lower part of the pipe about two inches beneath the Sur-face of ſome oyl of Turpentine (which, to make the effect the clearer, I ſome-times tinge deeply with Copper :) then removing my finger, the oyle being preſs'd againſt the immers'd Orifice with a greater force, then the weight of ſo little ſuſpended water could re-ſiſt, that oyle was impell'd into the lower part of the pipe to the height of near an inch ; and then again I ſtopp'd the upper Orifice of the pipe with my

finger,

finger, and thereby keeping both the Liquors fuſpended in it, I thruſt the pipe into a Glaſs full of water, three or four inches beneath the Surface of it; and then (for the Reaſon juſt now given) the water, upon the Removal of my finger, will preſs in at the lower Orifice of the pipe, and impell up the *Fig.* 17. oyle, till they come to ſuch a ſtation, as that expreſs'd in the annex'd Scheme : where $P Q$ is the water, newly impell'd up into the pipe, $Q R$ is the oyl, and $R S$ the water that was at firſt ſuck'd into the pipe. For in this ſtation, theſe three liquors do altogether as much gravitate upon the part P, as the incumbent water alone does upon the other parts of the imaginary ſuperficies $G H$; and yet the oyle, $R Q$, does not aſcend, becauſe the difluence of the water, $R S$, being hindred by the ſides of the pipe, its ſuperficies, $T S$, is higher then $A D$, the

Super-

Superficies of the reft of the water; by which means the incumbent water may be brought to have upon the upper part R of the oleous Cylinder, as great a preffure as that of the water, that endeavours to impel upwards the lower part Q of the fame fufpended Cylinder of oyle.

PARA-

PARADOX X.

That the cauſe of the Aſcenſion of water in Syphons, and of its flowing through them, may be explicated without having a recourſe to natures abhorrency of a Vacuum.

BOth Philoſophers and Mathematicians, having too generally confeſt themſelves, reduc'd to fly to a *fuga vacui*, for an account of the cauſe of the running of water and other Liquors through Syphons. And ev'n thoſe moderns, that admit a *Vacuum*, having (as far as I have met with) either left the Phænomenon unexplicated, or endeavour'd to explain it by diſputable Notions: I think the Curious much oblig'd to

Monſieur

Monfieur *Pafchal*, for having ingenioufly endeavour'd to fhew, That this difficult Probleme need not reduce us to have recourfe to a *fuga vacui*. And indeed his Explication of the motion of water in Syphons, feems to me fo confonant to Hydroftatical principles, that I think it not neceffary to alter any thing in it. But as for the experiment he propounds to juftifie his Ratiocination, I fear his Readers will fcarce be much invited to attempt it. For, befides that it requires a great quantity of Quickfilver; and a new kind of Syphon, 15 or 20 foot long; the Veffels of Quickfilver muft be plac'd 6 or 7 yards under water, that is, at fo great a depth, that I doubt whether men, that are not divers, will be able conveniently to obferve the progrefs of the Tryal.

Wherefore we will fubftitute a way, which may be try'd in a glafs Tube, not

two

two foot deep, by the help of another
peculiarly contriv'd glafs, to be prepar'd
by a skilful hand. Provide then a glafs
Tube *A B C D*, of a good widenefs,
and half a yard or more in depth; provide alfo a Syphon of two legs *F K*,
and *K G*, whereunto is joyn'd (at the
upper part of the Syphon) a pipe *E K*,
in fuch manner, as that the Cavity of
the pipe communicates with
the cavities of the fyphon; fo   *Fig.* 18.
that if you fhould pour in
water at *E*, it would run out at *F* and
G. To each of the two Legs of this
new Syphon, muft be ty'd with a ftring
a pipe of Glafs, *I* and *H*, feal'd at one
end, and open at the other; at which it
admits a good part of the leg of the
Syphon to which it is faftned, and
which leg muft reach a pretty way
beneath the Surface of the water,
wherewith the faid pipe is to be almoft
fill'd. But as one of thefe legs is longer
                                        then

then the other, fo the furface of the
water in the fufpended pipe, *I*, that is
faftned to the fhorter leg *K F*, muft be
higher (that is, nearer to *K* or *A B*)
then the furface of the water in the
pipe *H*, fufpended from the longer leg
KG; that (according to what is ufual in
Syphons) the water may run from a
higher veffel to a lower.

All things being thus provided ; and
the pipe *E K* being held, or otherwife
made faft that it may not be mov'd ;
you muft gently poure oyle of Turpentine into the Tube *A B C D,*
(which, if you have not much oyle,
you may before hand fill with water
till the liquor reach near the Bottom
of the fufpended pipes, as to the fuperficies *X Y*) till it reach higher then the
top of the Syphon *F K G*, (whofe
Orifice *E* you may, if you pleafe, in
the mean time clofe with your finger
or otherwife , and afterwards unftop)
and

and then the oyle preſſing upon the water will make it aſcend into the legs of the Syphon ; and paſs through it, out of the uppermoſt veſſel *J*, into the lowermoſt H ; and if the veſſel *J* were ſupply'd with water, the courſe of the water through the Syphon would continue longer, then here (by reaſon of the paucity of water) it can do.

Now in this Experiment we manifeſtly ſee the water made to take its courſe through the legs of a Syphon from a higher veſſel into a lower, and yet the top of the Syphon being perforated at K, the aire has free acceſs to each of the legs of it, through the hollow pipe *B K* which communicates with them both. So that, in our caſe, (where there is no danger of a *Vacuum*, though the water ſhould not run *through* the Syphon) the fear of a *Vacuum* cannot with any ſhew of Reaſon be pretended to be the cauſe of its running. Wherefore

fore we muſt ſeek out ſome other.

And it will not be very difficult to find, that 'tis partly the preſſure of the oyle, and partly the contrivance and ſituation of the veſſels ; if we will but conſider the matter ſomewhat more atentively. For the oyle, that reaches much higher then K, and conſequently then the leggs of the Syphon , preſſes upon the ſurface of the External water, in each of the ſuſpended pipes I and H. I ſay the *External water*, becauſe the oyle floating upon the water , and the Orifices of both the legs F and G be-ing immers'd under the water, the oyle has no acceſs to the cavity of either of thoſe legs. Wherefore, ſince the oyle gravitates upon the water without the legs , and not upon that within them, and ſince its height above the water is great enough to preſs up the water into the Cavity of the legs of the Sy-phon, and impel it as high as K, the

water

water muſt by that preſſure be made to aſcend.

And this raiſing of the water happening at firſt in both legs, (for the cauſe is in both the ſame) there will be a kind of conflict about *K* betwixt the two aſcending portions of water, and therefore we will now examine which muſt prevaile.

And if we conſider, That the preſſure, ſuſtein'd by the two parcels of water in the ſuſpended pipes *I* and *H*, depends upon the height of the oyle that preſſes upon them reſpectively ; it may ſeem (at the firſt view) That the water ſhould be driven out of the lower veſſel into the higher. For if we ſuppoſe that part of the ſhorter leg that is unimmers'd under water to be 6 Inches long, & the unimmers'd part of the longer leg to be ſeaven Inches ; becauſe the ſurface of the water in the veſſel *I*, is an Inch higher, then that of

the

the water in the veſſel *H*, it will follow,
That there is a greater preſſure upon
.the water, whereinto the longer leg is
dip'd, by the weight of an Inch of oyle:
ſo that that liquor being an inch higher
upon the ſurface of the water in the
pipe *H*, then upon that in the pipe *I*,
it ſeems that the water ought rather to
be impell'd from *H* towards *K*, then
from *I* towards *K*.

But then we muſt conſider, That,
though the deſcent of the water in the
leg *G*, be more reſiſted then that in
the other leg, by as much preſſure as
the weight of *an Inch of oyle* can amount
to ; Yet being longer by an Inch then
the water in the leg *F*, it tends downwards more ſtrongly by the weight of an
Inch of water, by which length it exceeds
the water in the oppoſite leg. So that
an inch of water being *(cæteris paribus)*
heavier then an Inch of oyle ; the water in the longer leg , notwithſtanding

N   the

the greater resistance of the external
oyle; has a stronger endeavour downwards, then has the water in the shorter
leg; though the descent of this be resisted but by a depth of oyl less by an Inch.
So that all things computed, the motion must be made towards that way
where the endeavour is most forcible;
and consequently the course of the water must be from the upper vessel, and
the shorter leg, into the longer leg, and
so into the lower vessel.

The application of this to what
happens in Syphons is obvious enough.
For, when once the water is brought
to run through a Syphon, the aire
(which is a fluid, and has some gravity,
and has no access into the cavity of the
Syphon,) must necessarily gravitate
upon the water whereinto the legs of
the Syphon are dip'd; and not upon
that which is within the Syphon : and
consequently, though the incumbent

aire

aire have somewhat a greater height
upon the water in the lower veſſel, then
upon that in the upper ; yet the gra-.
vitation it thereby exerciſes upon the
former more then upon the latter, being very inconſiderable, the water in
the longer leg much preponderating
(by reaſon of its length) the water in
the ſhorter leg, the efflux muſt be out
of that leg, and not out of the other.
And the preſſure of the External aire
being able to raiſe water (as we find by
ſucking Pumps) to a far greater height,
then that of the ſhorter leg of the Syphon ; the efflux will continue, for the
ſame reaſon, till the exhauſtion of the
water, or ſome other circumſtance, alter the caſe. But, if the legs of the Syphon ſhould exceed 34 or 35 foot of
perpendicular altitude ; the water
would not flow through it;
the preſſure of the exter- In the Phyſico-
nal aire being unable, (as periments.

N 2   has

has been elsewhere declared,) to
raise water to such a height. And
if a hole being made at the top of a
Syphon, that hole should be unstopp'd
while the water is running, the course
of it would presently cease. For, in
that case, the aire would gravitate upon the water, as well within as without the cavity of the Syphon; and so
the water in each leg would, by its own
weight, fall back into the vessel belonging to it.

But because this last circumstance,
though clearly deducible from Hydrostatical principles and Experiments,
has not, that I know of, been verified
by particular Tryals, I caus'd two
Syphons to be made, the one of Tin,
the other of Glass; each of which had,
at the upper part of the flexure, a
small round hole or socket, which I
could stop and unstop, at pleasure,
with the pulp of my finger. So that,
when

when the water was running through
the Syphon, in cafe I remov'd my
finger, the water would prefently fall,
partly into one of the fubjacent veffels,
and partly into the other. And if the
legs of the Syphon were fo unequal
in length, that the water in the one
had a far greater height (or depth)then
in the other ; there feem'd-to be,when
the liquor began to take its courfe
through the Syphon, fome light preffure from the external aire upon the
finger,wherewith I ftopp'd the Orifice
of the focket made at the flexure.

And on this occafion I will add,
what I more then once try'd ; to fhew,
at how very minute a paffage the preffure of the External aire may be communicated, to Bodyes fitted to receive
it. For, having for this purpofe ftopp'd
the orifice of one of the above mentioned Syphons, (infteed of doing it with
my finger,) with a piece of oyl'd paper,

N 3   care-

carefully faſtned with Cement to the
ſides of the ſocket ; I found, as I expected, that though hereby the Syphon
was ſo well clos'd, that the water ran freely through : yet, if I made
a hole with the point of a needle, the
aire would at ſo very little an orifice
inſinuate it ſelf into the cavity of the
Syphon, and, thereby gravitating as
well within as without, make the
water in the legs to fall down into the
veſſels. And *though*, if I held the point
of the needle in the hole I made, and
then caus'd one to ſuck at the
longer leg; this ſmall ſtopple, without any other help from my hand,
ſuffic'd to make the Syphon fit for uſe :
Yet if I remov'd the needle , the aire
would (not without ſome noiſe) preſently get in at the hole , and put a
final ſtop to the courſe of the water.
Nor was I able to take out the needle
and put it in again ſo nimbly, but that
the

the aire found time to get into the
Syphon; and, till the hole were again
ftopp'd, render it ufelefs, notwithftanding that the water was by
fuction endeavour'd to be fet a running.

PARA-

T

PARADOX XI.

That a folid Body, as ponderous as any yet known, though near the Top of the water it will finck by its own weight; yet if it be plac'd at a greater depth then that of twenty times its own thicknefs, it will not finck, if its defcent be not affifted by the weight of the incumbent water.

THis Paradox, having never been (that I know of) propos'd as yet by any, has feem'd fo little credible to thofe to whom I have mention'd it, (without excepting Mathematicians themfelves,) that I can fcarce hope it fhould be readily and generally received in this Illuftrious Company, upon
lefs

lefs clear Teſtimony, then that of Experience. And therefore, though (if I
miſtake not) ſome part of this propoſition may be plauſibly deduc'd by the
help of an Inſtrument ingeniouſly
thought upon by *Monſieur Paſchal* ;
Yet I ſhall have recourſe to my own
Method for the making of it out, for
theſe two Reaſons. The one, That
a great part of the Paradox muſt be
Explicated, as well as prov'd, by the
Doctrine already ſetled in this paper.
The other, That the Experiment propos'd by *Monſieur Paſchal*, being to be
done in a deep River, and requiring a
Tube 20 foot long, whoſe Bottome
muſt be fitted with a Braſs Cylinder,
made with an exactneſs, ſcarce (if at
all) to be hoped for from our Work-
men : If I ſhould build any thing on this
ſo difficult an Experiment, (which
himſelf does not affirm to have ever
been actually tryed,) I fear moſt men
would

would rather reject the Experiment
as a Chimærical thing, then receive for
its fake a Doctrine that appears to
them very Extravagant.

Let us then, to imploy in this cafe
also the method we have hitherto made
ufe of, Fill a Glafs veffel, *A B C D*,
almoft full of water ; only, in
regard that there is a great   *Fig.* 19.
depth of water requifite to fome Cir-
cumftances of the Experiment, This
laft muft not be fo fhallow as thofe hitherto imploy'd : but a deep Cylinder, or Tube feal'd at one end, whofe
depth muft be at leaft two or three
foot, though its breadth need not be a-
bove 2 or 3 Inches ; and, to keep it upright, it may be plac'd in a focket of
metal or wood, of a fize and weight
convenient for fuch a purpofe. This
Glafs being thus fitted in water, let
us fuppofe *E F*, to be a round and flat
piece of folid Brafs, having about an
Inch

Inch in Diameter, and a fourth or fixth
part of an inch in thicknefs. This Cylinder, being immers'd under water till
it be juft cover'd by the uppermoft
Surface of that Liquor, and being let
go, muft neceffarily fall downwards in
it; becaufe if we fuppofe the imaginary Superficies, G H, to pafs along the
Circle F, which is the lower part of
the Brafs Body, that metal being *in
fpecie* far heavier then water, the Brafs
that leans upon the part F, muft far
more gravitate upon the faid part F,
then the incumbent water does upon
any other part of the Superficies $G. H$;
and, confequently, the fubjacent water
at F will be thruft out of place by the
defcending Body. And becaufe that,
in what part foever of the water, not
exceeding nine times its thicknefs mea-
fured from the Top of the water A C,
the ponderous Body, E F, fhall happen to be; there will be ftill, by reason

ſon of the ſpecifick gravity of the Metal, a greater preſſure upon that part
of the imaginary Superficies that
paſſes along the bottome of the Body
on which the part F ſhall happen to
lean, then upon any other part of the
ſame imaginary Superficies; the Braſs
Body would ſtill deſcend by vertue of
its own weight, though it were not aſſiſted by the weight of the water that
is over it. But let us ſuppoſe it to
be plac'd under water on the deſign-
able plain J K; and let this plain, which
(as all other imaginary plains) is,
as well as the real Surface of the water, to be conceiv'd parallel to the
Horizon ; and let the depth or diſtance of this plaine, from the uppermoſt Surface of the water, be
(ſome what) above nine times the thickneſs of the Braſs Body : I ſay that, in
this caſe, the body would not deſcend,
if it were not preſs'd downwards by
the

the weight of the water it has over it.
For Brass being but a-
bout nine times * as
heavy as water of an e-
qual bulk to it, the Bo-
dy *E F* alone would
press upon the part *F*,
but as much as a Cy-
linder of water would,

: The word, about, is added, because indeed the Author, as he elfe-where delivers, did by exact scales find Brass to weigh between eight or nine times as much as water; but judg'd it needless to his pre-sent Argument, and in-convenient to take no-tice of the fraction.

which having an equal Basis were 8 or
9 times as high as the Brass is thick.
But now all the other parts of the I-
maginary surfaces, *I K*, being press'd
upon by the incumbent water, which
is as high above them as the newly
mention'd Cylinder of water would
be ; there is no reason why the part *F*
should be deprels'd, rather then any o-
ther part of the Superficies *J K* : But
because it is true, which we formerly
taught ; namely, that water retains its
gravity in water ; and that too, though
a body, heavier *in specie* then it, be plac'd
immed

immediately under it; it will necessarily happen, That in what part soever the solid body be plac'd, provided it be every way environ'd with the water, it must, for the Reason newly given, be made to move downwards, partly by its own weight, and partly by that of the incumbent water; and must continue to sinck, till it come to the bottom, or some other body that hinders its farther descent.

But in case the water above the solid body did not gravitate upon it, and thereby assist its descent; or, in case that the incumbent water were by some Artifice or other, so remov'd, That none of the lateral water (if I may so call it) could succeed in its place to lean upon the solid; then it will follow, from what we have newly shown, that the solid would be kept suspended. And in case it were plac'd much deeper in the water, as over against the

point

point *L* or *M* ; Then, if we conceive
the incumbent water to be remov'd or
fenc'd off from it, the preſſure of the
ſolid alone upon the part *F*, of the imaginary Superficies *L M*, being very
much inferior to that of the water upon the other parts of the ſame Surface,
the part *F* would be ſtrongly impell'd
upwards, by a force proportionate to
the difference of thoſe two preſſures.
And therefore, ſince I have found by
tryals, purpoſely made in ſcales mar-
vellouſly exact, and with refined Gold,
(purer then perhaps any that was ever
weighed in water) That Gold, though
much the ponderouſeſt of bodies yet
known in the world, is not full 20 times
as heavy as water of the ſame Bulk ;
I kept within compaſs (as well as imploy'd a round number, as they call it)
when I ſaid, That no body (yet known,)
how ponderous ſoever, will ſubſide in
water by its own weight alone, if it
were

were fo plac'd under water, that the depth of the water did above twenty times exceed the height of the Body; (not to mention here, that though gold and water being weigh'd in the aire, their proportion is above 19 to one, yet in the water, gold does, as other fincking bodies, loofe as much of its weight, as that of an equal bulk of water amounts too.)

I was faying juſt now, that in cafe the Brazen body were plac'd low e-enough beneath the Surface of the water, and kept from being deprefs'd by any incumbent water, it would be fupported by the fubjacent water. And this is that very thing that I am now to ſhew by an Experiment.

Let then the Brafs body *E F*, be the cover of a brafs Valve; *Fig.* 20. (as in the annexed figure:) and let the Valve be faſtned with fome ſtrong and clofe Cement

to

to a Glafs pipe, *O P*, (open at both
ends) and of a competent length and
widenefs. For then the Body, *E F*,
being the undermoft part of the Inftrument, and not fticking to any other
part of it, will fall by its own weight
if it be not fupported. Now then, tying a thred to a Button ꝗ, (that is
wont to be made in the middle of the
doors of Brafs valves) you muft, by
pulling that ftring ftreight and upwards, make the Body, *E F*, fhut the
orifice of the Valve, as clofe as you can;
(which is eafily and prefently done.)
Then thrufting the Valve under water,
to the depth of a foot or more ; the Cement and the fides of the Glafs, *O P*,
(which reaches far above the top of
the water *X Y*) will keep the water
from coming to beare upon the upper
part of the body *E F* ; and confequently the imaginary Surface, *V W*, (that
pafles by the lower part of the faid

O   body)

body) will, where it is contiguous thereunto, be pref'sd upon only by the proper weight of the body $E\,F$; but in
its other parts, by the much greater
weight of the incumbent water. So
that, though you let go the ftring, (that
held the body, $E\,F$, clofe to the reft
of the Inftrument) the faid body will
not at all finck, though there be nothing but water beneath it to fupport
it.

And to manifeft that 'tis onely the
preffure of the water, of a competent
depth, that keeps the folid fufpended;
if you flowly lift up the inftrument towards ($X\,T$) the top of the water;
you fhall find, that, though for a while
the parts of the Valve will continue u-
nited, as they were before; yet, when
once it is rais'd fo near the Surface, (as
between the plain $\mathcal{J}\,K$, and $X\,T$) that
the fingle weight of $E\,F$, upon the fubjacent part of the imaginary plain that
paffes

paſſes by it, is greater then the preſſure of the incumbent water upon other parts of the ſame plain; that Body, being no more ſupported as formerly, will fall down, and the water will get into the pipe, and aſcend therein, to the level of the External water.

But if, when the Valve is firſt thruſt under water, and before you let go the thred that keeps its parts together, you thruſt it down to a good depth, as to the Superficies $R S$: then, though you ſhould hang a conſiderable weight, as L, to the Valve $E F$, (as I am going to ſhew you a Tryal with a Maſſy Cylinder of ſtone broader then the Valve, and of divers inches in length) the ſurpluſage of preſſure on the other parts of the plain, $V W$, (now in $R S$) over and above what the weight of the body $E F$, and that of the Cylindrical ſtone, L, to boot, can amount to, on that part of the Surface vvhich is contigu-

ous

ous to the said body *E F*, will be great
enough to press so hard against the
lower part of the Valve, that its own
weight, though assisted with that of
the stone, will not be able to disjoyne
them.

By which (to note that by the way)
you may see, that though, when two
flat and polish'd marbles are joyn'd
together, we find it is impossible to se-
ver them without force ; we need not
have recourse to a *fuga vacui*, to Explicate the cause of their Cohæsion,
whilst they are environ'd by the Aire,
which is a Fluid not devoid of Gravity, and reaching above the Marbles no
body knows how high.

And to evince, That 'tis only such
a pressure of the water, as I have been
declaring, that causes the Cohæsion of
the parts of the Valve ; if you gently
lift it up towards the top of the water,
you will quickly find the Brass body,
                                    E F,

E F, drawn down by the stone *(L)* that hangs at it ; as you will perceive by the waters getting in between the parts of the Valve, and afcending into the pipe.

To which I shall only add, what you will quickly fee, That, in perfect Conformity to our Doctrine, the preffure of the body, *E F,* upon the subjacent water, being very much increafed by the weight of the stone that hangs at it, the Valve needs not, as before, be lifted up above the plain *J K,* to overcome the refistance of the water, being now enabled to do it before it is rais'd near fo high.

APPEN-

APPENDIX I.

Containing an Anſwer to ſeven Objecti-
ons, propos'd by a late Learned Writer,
to evince, that the upper parts of wa-
ter preſs not upon the lower.

AFter I had, this Morning, made
an end of reviewing the foregoing
papers, there came into my hands ſome
queſtions lately publiſh'd, among other
things, by a very recent Writer of Hydroſtaticks. In one of which Queſtions, the Learned Author ſtrongly de-
fends the contrary to what has there
been in ſome places prov'd, and divers
places ſuppos'd.

The Author of theſe Erotemata
aſſerts,

afferts, That, *in confiftent water, the upper parts do not gravitate or prefs upon the lower.*

And therefore, I think it will be neither ufelefs, nor improper, briefly to examine here the Arguments he produces. Not ufelefs; becaufe the Opinion he afferts, both is, and has long been, very generally receiv'd; and becaufe too, it is of fo great importance, that many of the Erroneous Tenets and Conclufions, of thofe that (whether profeffedly or incidentally) treat of Hydroftatical matters, are built upon it. And not *improper*; becaufe our Learned Author feems to have done his Reader the favour to fumme up into one page all the Arguments for his Opinions that are difperfedly to be found in his own or others mens Books. So that in anfwering thefe, we may hope to do much towards a fatisfactory Decifion of fo important

O 4
a

a Controverſie. And, after what we
have already deliver'd, our Anſwers
will be ſo ſeaſonable, that they will
not need to be long : The things they
are built on having been already made
out, in the reſpective places whereto the
Reader is referr'd.

Our Author then maintains,
that, in Conſiſtent water, the Superiour do not actually preſs the Inferiour parts, by the ſeven following
Arguments.

Object. 1. Sayes he, *Becauſe elſe
the inferiour parts of the water would be
more denſe then the Superior, ſince they
would be compreſs'd and condens'd by the
weight of them.*

Anſ. But if the Corpuſcles, whereof
water conſiſts, be ſuppos'd to be perfectly ſolid & hard; the inferior Corpuſcles
may be preſs'd upon by the weight of
the ſuperior, without being compreſs'd
or condens'd by them. As it would
happen

happen, if Diamond duſt were lay'd together in a tall heap : For though the
upper parts, being heavy and ſolid Corpuſcles , cannot be deny'd to lean
and preſs upon the lower ; yet theſe,
by reaſon of their Adamantine hard-
neſs, would not be thereby compreſs'd.
And 'tis poſſible too , that the Corpuſcles of water , though not ſo per-
feƈtly hard, but that they may a little
yield to an extream force, be ſolid e-
nough not to admit from ſuch a
weight, as that of the incumbent water,
(at leaſt in ſuch ſmall heights as obſervations are wont to be made in,) any
compreſſion, great enough to be ſen-
ſible ; As, beſides ſome Tryals I have
formerly mention'd in another place,
thoſe made in the preſence of this
Illuſtrious Company ſeem ſufficiently
to argue ; viz. That water is not ſen-
ſibly compreſſible - by an ordinary
force. And I find not, by thoſe that
make

make the Objection, that they ever took pains to try, whether in deep places of the Sea, the lower parts are not more condens'd then the upper : nor do I see any absurdity, that would follow from admitting them to be so.

Object. 2. Our Authors second Argument is, *Because Divers feel not, under water, the weight of the water that lyes upon them.*

Anf. But for Answer to this Argument, I shall content my self to make a reference to the ensuing Appendix, where this matter will be confidered at large ; and where, I hope, it will be made to appear, that the phænomenon may proceed, partly from the firm Texture of the Divers body, and partly from the nature of that preffure which is exercis'd againft bodyes immers'd in fluids ; which, in that cafe, (as to fenfe) preffes every where equally, againft all
the

the parts of the body, expos'd to their
Action.

Object. 3. The third Argument is,
*That ev'n the slightest Herbs growing at the
bottom of the water, and shooting up in it
to a good height, are not oppress'd or lay'd
by the incumbent water.*

Anf. But the Anſwer to that is eaſie,
out of the foregoing Doctrine. For the
Plants, we ſpeak of, ſuſtain not the
preſſure of the water above them by
their own ſtrength ; but by the help
of the preſſure of water that is beneath: which being it ſelf preſs'd by the
water that is (though not perpendicularly over it) ſuperior to it, preſſes them
upwards ſo forcibly, that if they were
not by their Roots, or otherwiſe
faſtned to the ground, they, being *in ſpecie* lighter then water,
would be buoy'd up to the top of the
water, and made to float ; as we often
ſee that weeds do, which ſtorms, or
other

other accidents have torn from their native foyle.

Objeɛt. 4. A fourth Objeɛtion is this, *That a heavy Body ty'd to a ſtring, and let down under water, is ſupported, and drawn out with as much eaſe, as it would be if it had no water incumbent on it; nay, with greater eaſe, becauſe heavy bodyes weigh leſs in water then out of it.*

Anſ. But an Account of this is eaſie to be rendred out of our Doɛtrine; For, though the water incumbent on the heavy body do really endeavour to make it ſinck lower, yet that endeavour is rendred ineffeɛtual, to that purpoſe, by the equal preſſure of the water upon all the other parts of the Imaginary ſurface, that is contiguous to the bottom of the immers'd body. And that preſſure upon the other parts of that ſuppos'd plain, being equal not only to the preſſure of the pillar of water,

ter, but to that pillar , and to the
weight of as much water as the immers'd body fills the place of ; it muſt
needs follow, That not only the hand
that fuſteins the body, ſhould not feel
the weight of the incumbent water,
but ſhould be able to lift up the Body
more eaſily in the water , then in the
aire. But though the preſſure of the
water incumbent on the ſtone can not,
for the reaſon aſſign'd, be felt in the caſe
propos'd; yet if you remove that water,
(as in the Experiment brought for the
proof of the laſt Paradox,) it will
quickly appear by the preſſure againſt
the lower part of the heavy body,
and its inability to deſcend by its own
weight, when it is any thing deep under
water; it will (I ſay) quickly appear,
by what will follow upon the abſence
of the Incumbent water, how great a
preſſure it exercis'd upon the ſtone
whilſt it lean'd on it.

Object.

Object. 5. The fifth Argument is propos'd in thefe words , *Becaufe a Bucket full of water , is lighter in the water, then out of it ; nor does weigh more when full within the water , then when empty out of it ; nay it weighs lefs, for the reafon newly affign'd* (in the fourth Objection ;) *therefore the water of the Bucket, becaufe it is within water, does not gravitate, nor confequently prefs downwards, either the Bucket, or the water under the Bucket.* This is the grand and obvious Experiment, upon which the Schools, and the generality of Writers, have very confidently built this Axiom: *That the Elements do not gravitate in their proper place* ; and particularly, that water weighs not (as they fpeak) in its own Element.

Anf. What they mean by proper or *natural place*, I fhall not ftand to examine, nor to enquire whether they can prove, that water or any other fub-
lunary

lunary body poffeffes any place, but upon this account, that the caufe of gravity, or fome other movent, enables it to
expel other contiguous Bodies (that are
lefs heavie or lefs moved,) out of the
place they poffefs'd before ; and gives
it an inceffant tendencie, or endeavour
towards the lowermoft parts of the
Earth.

But as to the Example propos'd, its
very eafie to give an account of it.
For fuppofe *A B C D*, to be a Well ;
wherein, by the ftring *E F*, the Bucket is fufpended under water, and has
its Bottom contiguous to the imaginary
plain *I K*. If now we fuppofe the Bucket
to confift only of wood, lighter then
water, it will not only not prefs upon
the hand that holds the Rope at *E*, but
will be buoyd up, till the upper parts of
the Bucket be above the top of the
water ; becaufe the wood, whereof the
Bucket is made, being lighter *in fpecie*

then

then water, the preſſure of the water
in the Bucket *G*, and the reſt of the
water incumbent on that, together
with the weight of the Bucket it ſelf,
muſt neceſſarily be unable to preſs the
part *H* ſo ſtrongly, as the other parts
of the imaginary plaine *I K* are preſs'd
by the weight of the meer water incumbent on them. But if, as tis uſual,
the Bucket conſiſts partly of wood,
partly of iron ; the Aggregate may often indeed be heavier then an equal
bulk of water : But then the hand, that
draws up the Bucket by the Rope
F E, ought not, according to our
Doctrine, to feel the weight of all the
Bucket, much leſs that of the water
contein'd in it. For though that aggregate of wood and iron, which we here
call the Bucket, be heavier then ſo
much water ; yet it tends not downwards with its whole weight, but only
with that ſurpluſage of weight, whereby

by it exceeds as much water as is e-
qual to it in Bulk ; which furplufage
is not wont to be very confiderable.
And as for the water in the Cavity ,
G, of the Bucket, there is no reafon
why it fhould at all load the hand at E,
though really the water both in the
Bucket and over it do tend downwards with their full weight ; becaufe
that the reft of the water, L I, and M K,
do full as ftrongly prefs upon the reft
of the imaginary Superficies I K, as
the Bucket and the incumbent water
do upon the part H : and confequently
the bottom of the Bucket is every
whit as ftrongly prefs'd upwards by the
weight of the water, upon all the o-
ther parts of the plain I K; as it tends
downwards, by virtue of the weight
of the Incumbent water, that is partly
in the Bucket, and partly above it;
and fo thefe preffures ballancing one
another , the hand that draws the

Rope at *E*, has no more to lift up then
the furplufage of weight, whereby
the empty Bucket exceeds the weight
of as much water as is equall in bulk
(I fay, not to the Bucket as 'tis a hollow
Inftrument, but) to the wood and iron
whereof the Bucket confifts.

And becaufe this Example of the
lightnefs of fil'd Buckets within the water has for fo many Ages gain'd credit
to, if it have not been the only ground
of, the affertion, That water weighs
not in its own Element, or in its proper
place ; I fhall add (though I can fcarfe
prefent it to fuch a *company* as this without fmiles) an Experiment that I made
to convince thofe, that were, through
unskilfulnefs or prejudice, indifpos'd to
admit the Hydroftatical account I have
been giving of the phænomenon. I
took then a round wooden Box, which
I fubftituted in the room of a Bucket ;
and (having fill'd it with melted
Butter,

Butter, into which, when it was congeal'd, some small bitts of lead were put, to make it a little heavier then so much water,) I caus'd a small string of twin'd silk to pass through two small holes, made in the opposite parts of the upper edge of the box, and to be suspended at one end of the beam of a pair of Gold-smiths Scales; and then putting it into a vessel full of water, till it was let down there, to what depth I pleas'd, it appear'd that not only the least endeavour of my hand would either support it, or transport to and fro in the water, or draw it up to the top of it; and this, whether the box were made use of, or whether the butter and lead alone, without the box, were suspended by the silken string: but (to evince, that it was not the strength of my hand, or the smallness of the immers'd body, that kept me from feeling any considerable resi-

stance)

ftance,) I caft fome grains into the fcale that hung at the other end of the above mention'd Beame, and prefently rais'd the Lead and Butter to the furface of the water. So that unlefs the Schoolmen will fay that the butter & lead were in their own Element; we muft be allow'd to think, that the eafie fuftentation, and elevation of the box, did not proceed from hence, That thofe bodyes weigh'd not becaufe they were in their natural place. And yet in this cafe, the effect is the fame with that which happens when a bucket is drawing out of a well.

And, to manifeft that 'twas the preffure of the water againft the lower part of the furface of our fufpended body, that made it fo eafie to be fupported in the water, or rais'd to the top of it; I fhall add, that though a few grains fuffic'd to bring the upper furface of the butter to the top of the water:

yet

yet afterwards there was a confiderable
weight requifite, to raife more & more
of its parts above the waters furface ;
& a confiderabler yet, to lift the whole
body quite out of the water. Which
is very confonant to our Doctrine. For,
fuppofe the bucket to be at the part
N, half in and half out of the water :
the hand or counterpoife, that fupports
it in that pofture, muft have a far greater ftrength then needed to fuftein it,
when it was quite under water ; becaufe that now the imaginary plain
P Q, paffing by the bottom of the
bucket, has on its other parts but a
little depth of water, as from L to P,
or M to Q, and confequently the bottom of the bucket, H, will fcarce be
prefs'd upwards above half as ftrongly
as when the bucket was quite under
water. And if it be raifed to O, & confequently quite out of the water ; that
liquor reaching no longer to the bottom

of

of the bucket, can no longer contribute to its supportation ; and therefore a weight not only equal, but somewhat superiour to the full weight of the bucket, and all that it contains, (being all suppos'd to be weighd in the aire,) will be neceflary to lift it clear out of the water.

But to dwell longer on this subject cannot but be tedious to those that have been any thing attentive to the former Difcourfes. I proceed therefore to our Authors fixth Argument, which is,

Object. 6. *That Horfe-hairs , which are held to be of the fame gravity with water, keep whatever place is given them in that Liquor; nor are deprefs'd by the weight of the fuper-incumbent water.*

Anfw. Whether the matter of fact be ftrictly and univerfally true, is fcarce worth the examining , efpecially fince we find the difference in point of fpecifick

cifick gravity, betwixt moſt Horſe-
haires, and moſt waters, to be inconſiderable enough. But the phænomenon,
ſuppoſing the truth of it, is very eaſily
explicable, according to the Doctrine
above deliver'd. For ſuppoſing in the
laſt Scheme the body, R, to be bulk for
bulk exactly equiponderant to water ;
'tis plain there is no reaſon why that
body ſhould preſs the part S, of the
imaginary Superficies $I K$, either
more or leſs then that part S would
be preſs'd, if, the body R being anni-
hilated or remov'd, it were ſucceed-
ed by a parcel of water of juſt the
ſame bulk and weight. And conſequently, though all the water directly
above the ſolid R do really lean upon
that body, and endeavour to depreſs it;
yet that endeavour being refiſted by an
equal and contrary endeavour, that
proceeds (as we have been but too often faine to declare) from the preſſure

P 4   exercis'd

exercis'd upon the other parts of the Superficies, *I K*, by the water incumbent on them; the body, *R*, will be neither deprefs'd nor rais'd. And its cafe being the fame in what part of the water foever it be plac'd, provided it be perfectly environ'd with that Liquor; it muft keep in the water (which in this whole Difcourfe we fuppofe to be Homogeneous as to gravity) the place you pleafe to give it.

And, (to add That on this occafion) though Mathematicians have hitherto contented themfelves to prove, that in cafe a Body could be found or provided, that were exactly equiponderant to water, it would retaine any affignable place in it; yet the Curiofity we had, to give an Experimental proof of this Truth, at length produc'd fome glafs Bubbles, which fome Gentlemen here prefent have not perhaps forgot, that were (by a dexterous hand we
employ'd

employ'd about it)fo exquifitely pois'd,
as, to the wonder of the Beholders, to
retain the places given them,fometimes
in the middle,fometimes near the top,&
fometimes near the bottom of thewater
(though that were Homogeneous) for a
great while, till fome change of confi-
ftence or gravity in the water, or fome
of its parts,made the bubble rife or fall.
  The Application of this, to what
has been objected concerning Horfe-
hairs, being too eafie to need to be in-
fifted on, there remains to be difpat-
ched our Authors feventh and laft Argument, which is this.
  Object. 7. *That, otherwife, all the*
inferiour parts of the water would be in
perpetual motion, and perpetually expell'd
by the Superior.
  Anfw. But if, by the inferior parts,
he means, fuch portions as are of any
confiderable bulk; the Anfwer newly
made to the laft objection (where we
                              fhew'd

fhew'd that the body, *R*, would retain
its place any where in the water, and
confequently near the bottome) will
fhew the invalidity of this Objection.
And unlefs we knew of what bigneffe
and fhape the Corpufcles of water are,
it would perhaps be to little purpofe
to difpute how far it may be granted,
or may be true in the particles that
water is made up of. Onely this I
fhall add, That, whereas this Learned
Authour mentions it as an abfurdity,
that the lower parts of water fhould
be in perpetual motion : And *Stevinus*
himfelf, in the beginning of his Hydroftatical Elements, feems to me to
fpeak fomewhat inconfiderately of this
matter ; and though, as I lately faid,
I allow fuch fenfible bodies, as thofe
whofe gravity in water Writers are
wont to difpute of, to be capable of
retaining their places in water, if they
be *in fpecie* equiponderant to it : Yet I
am

am fo far from thinking it abfurd, that
the inferiour Corpufcles of water
fhould be perpetually in motion; that
I fee not how otherwife they could
conftitute a Fluid body, That reftlefs
Motion of their parts, being one of the
generaleft Attributes of Liquors; and
being, in water, though not immediately to be *feen*, yet to be eafily *difcover'd*
by its Effects: As, when Salt, be-
ing caft into water, the aqueous parts
that are contiguous to it, and confequently near to the bottom, do foon
carry up many of the faline ones, to
the very top of the water; where, after a while, they are wont to difclofe
themfelves in little floating grains of a
Cubical fhape.

But, of this reftlefs motion of the
parts of Liquors having profeffedly
treated elfewhere already;
I fhall add nothing at pre- *In the Hiftory of*
fluidity & firmnefs
fent : But rather take

notice

notice of what our Authour subjoyns to the last of his Arguments, (as the Grand thing which they suppose) in these words, *Ratio porro, a priori, hujus sententiæ videtur esse, quia res non dicitur gravitare nisi quatenus habet infra se Corpus levius se in specie.* The errorniousness of which conceit, if I should now go about solemnly to evince; I as well fear it would be tedious, as I hope it will be needless to those, that have not forgot what may concern this subject in the former part of the now at length finish'd discourse; and especially where I mention those Experiments, which show, That neither a stone, nor Gold it self, when plac'd deep under water, would sinck in it, if the Superiour water, that gravitates on it, did not contribute to its depression.

APPENDIX II.

Concerning the Reaſon why Divers, and others who deſcend to the Bottome of the Sea, are not oppreſs'd by the weight of the incumbent water.

AMongſt the difficulties that belong to the Hydroſtaticks, there is one which is ſo noble, and which does ſtill ſo much both exerciſe and poſe the wits of the Curious, That perchance it will not be unacceptable, if to the former Experiments we add, by way of Appendix, one that may conduce to the ſolving of this difficult problem; *viz.* Why men, deep under water, feel no inconvenience by the preſſure of ſo great

a

a weight of water as they are plac'd under ?

The common Anſwer of Philoſo-phers and other Writers to this puzling Queſtion, is, That the Elements do not gravitate in their own proper pla-ces; and ſo, water in particular has no gravitation upon water, nor conſequent-ly upon bodies every way ſurrounded with water. But that this Solution is not to be admitted, may be eaſily ga-ther'd from our proofs of the firſt Pa-radox, and from divers other particu-lars, applicable to the ſame purpoſe, that may be met with in the forego-ing papers.

A famous VVriter, and, for ought I know, the Recenteſt (except *Monſieur Paſchal*) that has treated of Hydro-ſtaticks, having rendred this Reaſon of the Phænomenon.

[*The Superior parts of conſiſtent water* (as he ſpeaks) *preſs not the inferior, un-*
leſs

lefs beneath the inferior there be a Body lighter in fpecie then water; and therefore, fince a humane Body is heavier in fpecie then water , it is not prefs'd by the incumbent water, becaufe this does not endeavor to be beneath a humane Body.] He fubjoyns, contrary to his Cuftome, this confident Epiphonema , *Qui aliam caufam hujus rei affignant, errant & alios decipiunt.*

But, by his favour, notwithftanding this confidence, I fhall not fcruple to feek another Reafon of the Phænomenon. For I have abundantly prov'd,that (contrary to the Affertion on which his Explication is built) the upper parts of water prefs againft the lower, whether a body heavier or lighter *in fpecie* then water be underneath the lower. And, the contrary of which being the πρῶτον ψεῦδος in this Controverfie , perhaps the matter may be fomewhat cleared, by mentioning here a diftincti-
on,

on, which I fometimes make ufe of. I
confider then a body may be faid to
gravitate upon another body in two
fenfes. For fometimes it actually fincks
into, or gets beneath the body that was
under it, as a fincking ftone gravitates
upon water, and which I call Præva-
lent,or fuccefsful Gravitation;& fometimes it does not actually, at leaft not
vifibly defcend, but only exercifes its
gravitation by preffing againft the fubjacent body that hinders its defcent ; as
when a VVoman carries a Paile of water on her head, though the weight do
not actually get nearer the Center of
the Earth ; yet actually preffes with
its whole gravity upon the Womans
head, and back, and other fubjacent
parts that hinder its actual defcent ;
and according to this Doctrine I cannot admit our Authors reafoning, that
becaufe a mans body is bulk for bulk
heavier then water, therefore the water

ter does not endeavour to place its self
beneath it. For water, being a heavy
body, derives from the caufe of its gravity, (what ever that be) an inceflant
endeavour towards the Center of the
Earth; nor is there any Reafon, why
its happening to be incumbent on a
body heavier *in fpecie* then it felf,fhould
deftroy that endeavour. And therefore, though it may be faid that the
water does not endeavour to place it
felf beneath a humane body, becaufe
indeed an inani .ate Liquor cannot properly be faid to act for this or any o-
ther end; yet the water being a heavy
body, tends continually towards the
lower part of the Earth; and therefore will get beneath any body that is
plac'd betwixt it and that, (without
regard whether the inferior body be
heavier or lighter *in fpecie* then it felf)
as far as the degree of its gravity will
enable it; nor would it ever reft, till
it

it have reach'd the lowermoſt parts of
the Earth, if the greater ponderouſneſs
of the earth and other heavy bodies did
not hinder , (not its endeavour downwards, nor its preſſure upon ſubjacent
bodies, but only) its actual deſcent.

This Learned Author himſelf tells
us, (as well as *Stevinus*, and others, that
have written of the Hydroſtaticks, una-
nimouſly teach,) that if the bottom of
a veſſel be parallel to the Horizon, the
weight of water, that reſts upon it, is e-
qual to a pillar of water, having that
bottome for its Baſis, and for its height
a perpendicular reaching thence to the
uppermoſt Surface of the water. Nor
is it reaſonable to conceive that there
will be any difference in this preſſure
of the incumbent water, whether the
bottom be of Deale that will ſwimme,
or of Box that will ſinck in water; or
to ſpeak more generally, whether it be
of Wood, *in ſpecie* lighter then water,

or

or of Copper, or some other Metal,
that is *in specie* heavier then it. And
since water, being not a solid Body,
but a fluid, consists (as other fluids)
of innumerable Corpuscles , that ,
though extreamly minute, have their
own sizes and figures; And since the
preſſure of water upon the bottom of
a veſſel is proportionate to its perpendicular height over the bottom ; 'Tis
manifeſt, that the upper Corpuscles
preſs the bottom as well as the lower ;
which, since they cannot do immediately, they muſt do by preſſing the
intermediate ones. And I have already ſhown (diſcourſing one of the
former Paradoxes,) that the Superi-
or parts of water do not onely preſſe
thoſe that are directly under them, but
communicate a preſſure to thoſe that
are aſide of them, and at a diſtance
from them.

And if it be objected, That water

endea-

endeavours to get beneath a Bottome
of Glafs Veffels, or other bodies heavier *in fpecie* then its felf, becaufe under that bottome there is aire, which
is a body lighter *in fpecie* then water :
I fay, that this is precarious; for the
indifputable gravity of the water is a-
lone fufficient to make it always tend
downwards, (though it cannot always
move downwards) what ever body
be beneath it. And who can affure the
makers of this Objection, That there
are not beneath even the bottome of
Rivers, or of the Sea, (where yet they
fay water is confiftent, and refts as
in its own place,) vaft fpaces repleni-
fhed but with aire, fumes, or fire, or
fome other body lighter then water ?
For, (not to mention that the Carte-
fians take the Earth we tread on, to be
but a thin Cruft of the Terreftrial
Globe, whofe infide, as farre as the
Center, is replenifh'd with a fubtle fluid
matter,

matter, like that whereof the Sunne confifts.) We know that in fome pla-ces, as particularly at a Famous Coal-mine in *Scotland*, there are great Ca-vities that reach a good way under that ground that ferves there for a bot-tome to the Sea : So that, for ought thefe Objeɛtors know, ev'n according to their own Doɛtrine, the water ev'n in the Sea, may endeavour to get be-neath a body heavier *in fpecie* then it felf.

But, for my part, I cannot but think, that, to imagine the water knows, whether or no there be aire or fome lighter body then its felf beneath the body it leans on, and the fuperior parts do accordingly exercife or fuf-pend their preffure upon the inferior ; is to forget that it is a heavy Liquor, and an inanimate Body.

Another Solution there is of this Hydroftatical problem, we have been

dif-

difcourfing of, which I met with in a Printed Letter of Monfieur *Des Cartes*, in thefe terms.

Je ne me, &c. *I remember not what reafon 'tis that* Stevinus *gives, why one feels not the weight of water, when one is under it : but the true one is, that there can no more of water gravitate upon the body that is in it, or under it, then as much water as could defcend in cafe that body left its place.*

Second Tome lettre 32.

Fig. 22.

Thus for Example : If there were a Man in the Barrel, B, that fhould with his Body fo ftop the hole, A, as to hinder the waters getting out, he would feel upon himfelf the weight of the whole Cylinder of water, A B C, of which I fuppofe the Bafis to be equal to the hole A ; for as much as if he funck down through the hole, all the Cylinder of water would defcend too, but if he be a little higher, as about B, fo that he does no longer hinder the water from running

out

out at the hole **A**, he ought not to feel any
weight of the water which is over him, betwixt **B** and **C**, because if he should descend toward **A**, that water would not descend with him, but contrarywise a part of
the water which is beneath him towards
A, of equal bulk to his Body, would afcend into its place : so that in stead of feeling the water to press him from the *T*op
downward, he ought to feel that it buoys
him upward from the bottome ; which by
Experience we see.

Thus far this subtil Philosopher :
for whose Ratiocinations though I am
wont to have much respect, yet I
must take the liberty to confess my
self unsatisfy'd with this. For have-
ing already sufficiently prov'd, That
the upper parts of water press the
lower, and the bodies plac'd beneath
them, whether such bodies be lighter
in specie then water or heavier ; we
have subverted the Foundation, upon

which

which Monfieur *Des Cartes*'s ingenious, though unfatisfactory, Explication is built. And yet I fhall add *ex abundanti*, That fuppofing what he fayes, That in cafe the folid *B* fhould defcend towards *A*, the incumbent water would not defcend with it, but a part of the fubjacent water, equal in bulk to the folid, would afcend, and fucceed in its room ; yet that is but accidental, by reafon of the fteinchnefs and fulnefs of the Veffel. And though indeed the Superior water cannot actually defcend upon the depreffion of the folid at *B*, if, at the fame time while that body defcends, an equal bulk of water fucceeds in its place : Yet both the folid about *C*, and the water that fucceeds it, do, in their turns, hinder the defcent of the Superior water ; which therefore muft gravitate upon which foever of the two it be that actually comes to be

plac'd

plac'd directly under it, if there be
nothing, before the difplacing of the
folid, capable to take away the natural gravity, upon whofe account the
water, over *B* and *C*, does inceffantly
tend downwards. And though Monfieur *Des Cartes* does not fo clearly exprefs himfelfe, whether he fuppofes
the hole at *A* to be ftopp'd with fome
other body, when the folid is plac'd a-
bout *B* : yet, becaufe he is wont to
fpeak confiftently, I prefume he means,
that when the folid is remov'd to *B*,
the hole at *A* is otherwife fufficiently ftopp'd ; I fay then, that the
reafon why the folid, which, whilft at *A*,
fuftain'd a great preffure from the incumbent water, feels not the weight
of it, when plac'd at *B*, is not that
which *Monfieur des Cartes* gives, but
this, That the folid being environ'd
with water, the fubjacent water does (as
we have often had occafion to manifeft)
prefs

prefs it upwards, full as ftrongly (and
fomewhat more) as the weight of the
incumbent water prefles it downwards;
So that a mans body, in ftead of fincking, would be buoy'd up ; if, as it is a
little heavier, it were a little lighter *in*
fpecie then water. Whereas, when the
folid was that alone which cover'd and
ftop'd the hole, there was a manifeft
Reafon why it fhould be forcibly
thruft downwards by the weight of the
incumbent water *B C*. For, in that cafe,
there was no water underneath it at *A*,
to fupport the folid ; and, by its preffure upward, to enable it to refift fo
great a weight.

And this, (to hint that upon the by)
may perchance help us to guefs at the
reafon of what Geographers relate of
the Lake *Afphaltites* in *Judæa*, (in cafe
the matter of fact be true,) That this
dead Sea (as they alfo call it) will not
fuffer any living creature to finck in
it.

it. For the Body of a Man (and for ought we know of other Animals,) is not much heavier *in specie* then common fresh water : Now if in this Lake (that stands where *Sodom* and *Gomorrah* did, before those impious Regions were destroy'd by fire from Heaven,) we suppose, (which the nature of the Soyle, and the Sacred Story makes probable enough) That the water abounds with Saline or Sulphurous Corpuscles ; (the former helping the later to associate with the water, as we see in sope consisting of salt and oyle, and in Chymical mixtures of Alcalis and Brimstone dissoluble in water) the Liquor may have its gravity so augmented, as to become heavier *in specie* then the body of an animal. For I have learned of a Light Swimmer, that he could hardly begin to Dive in salt water, though he easily could in fresh. And 'tis not difficult to make a Brine

or

or *Lixivium* (which are but Solutions of salt in water,) heavy enough to keep up an egg from sincking. And, not only barely by diffolving a metalline body in a saline *Menstruum*, without otherwise thickning the Liquor, I have brought solid pieces of Amber it self to swim upon it : but I have try'd that certain saline Solutions, which I elsewhere mention ; nay, and a distill'd Liquor, (I us'd desteam'd oyle of Vitriol) without any thing diffolv'd in it, would do the same thing ; by reason of the numerous, though minute, Corpuscles of salt and sulphur, that it abounds with.

There remains but one solution more of our Hydrostatical probleme, that I think worth mentioning, and that is given by the Learned *Stevinus* in these words,

Omni pressu quo Corpus dolore afficitur, pars aliqua Corporis luxatur; sed isto

pressu

preſſu nulla Corporis pars luxatur, iſto igi-
tur preſſu Corpus dolore nullo afficitur.
Aſſumptio ſyllogiſmi manifeſta eſt, nam ſi
pars aliqua, ut caro, ſanguis, humor, aut
quodlibet deniq; mem-
brum luxaretur, in a-
lium locum concedat
neceſſe eſſet : atqui lo-
cus ille non eſt extra
Corpus ; cum aqua un-
diquaque æquali preſſu
circumfuſa ſit·(quod
vero pars ima, per 11.
propoſitionem Hydro-
ſtaticorum, paulo va-
lidius prematur ſupe-
riori, id hoc caſu nul-
lius momenti eſt, quia
tantula differentia par-
tem nullam ſua ſede
dimovere poteſt) neque
item intra ipſumCorpus
concedit, cum iſtic Cor-
pore omnia oppleta

Stevinus Hydroſtat. Lib. 5.
pag. 149.

Sed Exemplo clarius ita
intelliges,eſto ABCD aqua,
cujus fundum
D C, in quo fo-   Fig. 23.
ramen E ha-
beat Epiſtomiam ſibi inſer-
tum, cui Dorſo incumbat
HomoF,Quæ cum ita ſint,ab
aquæ pondere ipſi inſidente
nulla pars Corporis luxari
poterit, cum aqua,ut dictum
eſt, undiquaque æqualiter
urgeat.
Si vero ejus veritatem
explorare libeat, eximito
Epiſtomium, tumque tergum
nulla re fultumſuſtinebitur,
ut in locis cæteris, ideoque
iſtic tanto preſſu afficietur,
quantus tertio exemplo ſe-
cunda propoſitionis hujus
demonſtratus eſt : vid.
quantam efficit columna a-
quea cujus Baſis ſit fora-
men E. altitudo autem ea-
dem qua aquæ ipſi inſiden-
tis. Quo exemplo propoſiti
veritas manifeſte decla-
ratur.

ſint,

sint, unde singulæ partes singulis partibus
æqualiter resistunt, namque aqua undiqua-
que eadem ratione Corpus totum circum-
stat. Quare cum locus is nec intra, nec ex-
tra Corpus sit ; absurdum, imo impossibile
fuerit, partem ullam suo loco emoveri, ideo-
que nec Corpus hic afficitur dolore.

This Solution of *Stevinus*, I esteem
preferrible by farr, to those that are
wont to be given of this difficult Probleme : But yet, the Phænomenon
seems to me to have still somewhat in
it of strange. 'Tis true, that if the Question were only that which some put,
viz. Why the body of a Diver, when
it is near the bottom of the Sea, is not
pres'd down by so vast a weight of
water, as is incumbent on it ; It might
be rationally answer'd, That the
weight of so much water, as leans upon the body, is not sustein'd by the
force of the body it self, but by that
of the water which is under it. For,
by

by the Experiments and Explications,
we have annexed to fome of the foregoing Paradoxes, it appears, That the fubjacent water, by its preffure upwards,
is able, not only to fupport the weight
of the incumbent water, but fo far to
exceed it, that it would not only fupport the immers'd body, and the incumbent water, but buoy up the body,
if it were never fo little lighter *in fpecie*
then water. And as for what *Stevinus*
infinuates, That, when the water
preffes the body every way, that preffure is not felt, though it would be,
in cafe it prefs'd upon fome parts, and
not upon others; I am of the fame o-
pinion too; and, to prove it, fhall not
make ufe of the example he propofes,
in the words immediately following
thofe of his, I juft now recited: (For I
doubt, that example is rather a fup-
pofition, then a try'd thing;) but by
an Experiment which may be eafily
made,

made, and has diverſe times been ſo, in our Pneumatical Engine. For, though the aire be a heavy fluid, and though, whilſt it uniformely preſſes the whole ſuperficies of the body, we feel not the preſſure of it. And though, for this reaſon, you may lay the palm of your hand upon the open orifice of a ſmall braſs Cylinder, apply'd to the Engine inſtead of a Receiver, without any hurt; Yet when, by pumping, the aire that was before under the palm of your hand, is withdrawn, and conſequently can no longer help to ſupport your hand, a-gainſt the preſſure of the external and incumbent aire ; the external aire will lean ſo heavy upon the back of your hand , that you will imagine ſome pon-derous weight is lay'd upon it. And I remember by ſuch an Experiment, I have not onely had my hand put to much pain, but have had the back of it ſo bent downward, as if it were going to be broken. But

But though such confiderations, as
thefe, may much leffen the difficulty
of our phænomenon,whofe caufe is in-
quired into; Yet ftill it feems fomewhat odd to me, That (fince 'tis evident from the nature of the thing , and
by *Stevinus*'s his confeffion, that there
is a vaft preffure of water againft every
part of the body,whofe endeavour tends
inward,) fo exceedingly forcible a
preffure, (which thrufts, for inftance,
the Mufcles of the Arms and Thighs
againft the Bones, the Skin and Flefh
of the Thorax againft the Ribs,)
fhould not put the Diver to any fenfible pain ; As I find not (by one that
I examin'd) that it dos ; (Though this
man told me , he ftay'd a good while
at the depth of betwixt 80 and 100
foot under the Sea water, which is
heavier then frefh water ;) For, that
which *Stevinus*'s Explication will only
fhowis, That there muft be no mani-
R·   feft

feſt diſlocation of the greater parts of
the Body ; whereas the bare compreſſion of two ſmall parts, one againſt
another, is ſufficient to produce a ſenſe
of pain.

But it ſeems, the Texture of the
bodyes of Animals is better able to
reſiſt the preſſure of an every way
ambient fluid, then, if we were not
taught by experience, we ſhould imagine. And therefore, to ſatisfie thoſe
that (ſecluding the Queſtion about the
ſenſe of pain,) think it an abundantly
ſufficient Argument, (to prove, that
bodyes immers'd under water, are not
compreſs'd by it,) That *Divers* are
not oppreſs'd, and ev'n cruſh'd by ſo
vaſt a load of water, (amounting, by
Stevinus's computation, to many
thouſands of pounds) as is incumbent
on them. We will add, that though an
Experiment, propos'd by Monſieur
Paſchal to this purpoſe, were ſuch,
that

that at firft fight I faid that it would
not fucceed, (and was not upon tryal
miftaken in my conjecture;) yet it
gave me the occafion to make another,
which will, I hope, fully make out the
thing I defign'd it for.

The Ingenious Monfieur *Pafchal*
would perfwade his Readers, that if
into a glafs Veffel, with luke-warm
water in it, you caft a flie; and, by a
Rammer, forcibly prefs that water,
you fhall not be able to kill, or hurt the
flie. VVhich, fays he, will live as well,
and walk up and down as lively, in
luke-warme water, as in the aire. But,
upon tryal with a ftrong flie, the Animal was (as we expected,) prefently
drowned, and fo made moyelefs, by the
luke-warm water.

Wherefore we fubftituted another
Experiment, that we knew would not
only fucceed, (as you will prefently fee
it will do,) but teach us how great a

R 2 preffure

preſſure the included Animal muſt have been expos'd to. VVe took then a ſomewhat ſlender Cylin-

Fig. 24. drical pipe of Glaſs, ſeal'd at one end, and open at the other ; and to this we fitted a Rammer, which (by the help of ſome thongs of ſoft leather, that were carefully wound about it) did ſo exactly fill the pipe that it could not eaſily be mov'd to and fro ; and would ſuffer neither water, nor aire, to get by betwixt it, and the internal ſurface of the Glaſs. VVe alſo provided ſome ſmall Tad-poles (or *Gyrini*) about an Inch long or leſs ; which ſort of Animals we made choice of before any other, partly becauſe they could, by reaſon of their ſmalneſs, ſwim freely to & fro in ſo little water as our pipe contain'd ; & partly becauſe thoſe Creatures, being as yet but in their Infancy, were more, tender, and, conſequently, far

more

more expos'd to be injur'd by compreffion, then other Animals of the
fame Bulk, but come to their full age
and growth, would be, (as indeed fuch
young Tad-poles are fo foft and tender, that they feem, in comparifon to the
bigger fort of flies, to be but organiz'd
Gelly.) One of thefe Tadpoles being
put into the water, and fome Inches
of aire being left in the pipe, for the ufe
anon to be mention'd ; the water and
aire, and confequently the Tadpole,
were by the intrufion of the plug or
rammer, with as great a force as a man
was able to imploy, violently comprefs'd ; and yet, though the Tadpole
feem'd to be comprefs'd into a little
lefs Bulk then it was of before, it
fwom freely up and down the water,
without forbearing fometimes to afcend
to the very top, though the Inftrument were held perpendicular to the
Horizon. Nor did it clearly appear to

us,

us, That the little Animal was injur'd
by this compreffion ; and moſt manifeſt it is, he was not crufh'd to death,
or fenfibly hurt by it.

And having repeated this Experiment feveral times,&with Tadpoles of
differing ages;we may,I prefume,fafely
conclude, That the Texture of Animals is fo ſtrong, that,though water be
allowed to weigh upon water, yet a
Diver ought not to be oppreſt by It:
Since, whether or no water weighs in
water, 'tis manifeſt that in our Experiment, the water, and confequently the
Tadpole, was very forcibly by an External Agent comprefs'd betwixt the
violently condens'd aire, and the rammer. And, by the notice we took of
the quantity of aire before the com-
preflion began, and that to which it
was reduc'd by compreffion ; The mo-
derateſt eſtimate we could make, was,
That it was reduc'd into an *eighth*, or

tenth

tenth part of it's former space ; and so (according to what we have elswhere prov'd) the preſſure that was upon the aire, (and conſequently upon the water , and the included Tadpole,) was as great as that of a Cylinder of water of above 200 if not 300 foot high. And yet all this weight being unable to oppreſs, or ſo much as manifeſtly to hurt, the tender Tadpole (which a very ſmall weight would ſuffice to have cruſh'd, if it preſt only upon one part of it, and not upon the other) we may thence learn the Truth of what we have been endeavouring to evince : That though water be allowed to preſs againſt water, and all immers'd Bodys ; yet a *Diver* may very well remaine unoppreſs'd at a great depth under water, as long as the preſſure of it is uniforme againſt all the parts expos'd thereunto.